Not a Chance

Among Other Books by the Author

Before the Face of God. 4 vols.
Choosing My Religion
Chosen by God
Doubt and Assurance
Faith Alone
The Glory of Christ
Grace Unknown
The Holiness of God
The Invisible Hand
Knowing Scripture
The Last Days according to Jesus
Lifeviews
The Mystery of the Holy Spirit
Not a Chance
Now, That's a Good Question!
Pleasing God
Renewing Your Mind
The Soul's Quest for God
Surprised by Suffering
Ultimate Issues
Willing to Believe

Not a Chance

The Myth of Chance in Modern Science and Cosmology

R. C. Sproul

Baker Books

A Division of Baker Book House Co
Grand Rapids, Michigan 49516

Published by Baker Books
a division of Baker Book House Company
P.O. Box 6287, Grand Rapids, MI 49516-6287

Third printing, July 2002

Printed in the United States of America

Library of Congress Cataloging-in-Publication Data

Sproul, R. C. (Robert Charles), 1939–
    Not a chance : the myth of chance in modern science and cosmology / R. C. Sproul.
        p.   cm.
    Includes bibliographical references and index.
    ISBN 0-8010-5852-X
        1. Serendipity in science. 2. Chance. 3. Science—Philosophy.
    4. Cosmology. I. Title.
    Q172.5.S47S67   1994
    123'.3—dc20   94-26298

Photographs of Bohr and Heisenberg courtesy of the Bettmann Archive; of Einstein, Jaki, and Russell, Religious News Service. Portraits of Copernicus, Descartes, Galileo, Hume, Kant, Kepler, Locke, Ptolemy, and Voltaire courtesy of North Wind Picture Archives. Paintings by Magritte courtesy of Artists Rights Society, U.S.A. representative of C. Herscovici (Brussels), representative of the estate of René Magritte. Copernicus's model of the universe is from his *De Revolutionibus orbium coelestium* (1543); Kepler's model of the universe, from his *Mysterium cosmographicum* (1621); and Kepler's diagram of planetary orbits, from his *Harmonices mundi* (1619).

For current information about all releases from Baker Book House, visit our web site:
                http://www.bakerbooks.com

For information about Ligonier Ministeries and the teaching ministry of R. C. Sproul, visit Ligonier's web site:
                http://www.gospelcom.net/ligonier

To

Stanley L. Jaki

Contents

Illustrations
and Figures

Illustrations

Figures

Preface

In his controversial book *Worlds in Collision*, which once piqued the curiosity of Albert Einstein, Immanuel Velikovsky studied the mythology of ancient cultures in search of clues for prescientific information about astronomical perturbations and catastrophic cosmic upheavals. Velikovsky did not regard ancient myths as an exercise in sober historical narrative. He viewed mythology as fanciful, creative, imaginative attempts to explain the unknown powerful forces that impact human life. In a word, what we don't understand we tend to explain in terms of myths.[1]

We have a tendency in our day to think of mythology as a literary enterprise of primitive, ignorant, prescientific cultures. This tendency errs in two directions. On the one hand it is the nadir of arrogance for us to assume that ancient civilizations were primitive, ignorant, or prescientific. The Egyptians, Chinese, Babylonians, Romans, and Greeks, for example, were anything but primitive or ignorant. They all achieved extraordinary levels of scientific advancement. Yes, they had mythology; but they had their serious science as well.

The second error is to relegate mythology to the past, making it an addiction practiced only by pre-modern cultures. On the contrary, mythical approaches to life and learning persist in every culture. Mythology continues to intrude in the arena of religion. It is commonplace in the superstitions that abound among athletes in professional sports. It is found in a host of medicinal home remedies that are often classified under the rubric of "old wives' tales."

Mythology also intrudes into the realm of science. Uncritically accepted hypotheses and theories of the past die a slow and reluctant death. We have seen the resistance the church has displayed against new advances in scientific knowledge, the Galileo episode being the most famous. But it is not only churchmen who offer resistance. Even in Galileo's day opposition to him was heavily laden by scientists whose pet theories and accepted traditions were crumbling under the weight of new empirical evidence.

One myth that has found its way into modern thought and is entrenched in some circles is the myth of chance. In this myth the word *chance* itself undergoes an evolution and takes on new meaning. Where the word was once largely restricted to describing mathematical probability quotients, it took on a broader application to include far more than probabilities or coincidences. It has been used as a word to describe either the absence of cause or even a causal power itself. Mortimer Adler notes this new usage: "There is still a third sense of 'chance' in which it means that which happens totally without cause—the absolutely spontaneous or fortuitous."

With the elevation of chance to the level of a real force, the myth serves to undergird a chaos view of reality. Buttressed by inferences drawn from quantum theory, the idea that reality is irrational rather than coherent gained popularity.

James Gleick, in his book *Chaos: Making a New Science*,[2] describes the new shift away from chaos to new paradigms that seek the coherence underlying the surface appearance of chaos. Part of the struggle of science is the information explosion. As data proliferate, strain is put on old paradigms to accommodate them. The Hubble Space Telescope stretches our reach ever deeper into space, adding new information to the "far" of the universe. New levels of sophistication in microscopes pushes the horizons of the near and the small beyond former limits. As we probe the seemingly infinite and the infinitesimal, we are left with aching paradigms stretched to the breaking point.

The Enlightenment dream of discovering the "logic of the facts" has become a nightmare for many. Some have responded by abandoning logic altogether. It is when logic is negotiated or abandoned that myth is given fresh impetus. The twin enemies of mythology are logic and empirical data, the chief weapons of true science. If either weapon is neutralized, mythology is free to run wild.

This book is an effort to explore and critique the role chance has been given in recent cosmology. It may be viewed as a diatribe against chance. It is my purpose to show that it is logically impossible to ascribe *any power* to chance whatsoever.

It is not merely a parlor game of logic. There is something huge at stake: the very integrity, indeed the very possibility of science.

Diatribes may represent the unbridled ravings of fools. They may also represent the serious protests of the learned. I hope this work proves to be more of the latter than of the former.

R. C. Sproul
Orlando
Advent 1993

The Soft Pillow

"[Chance] has become for me a soft pillow like the one which . . . only ignorance and disinterest can provide, but this is a scientific pillow."

•

Pierre Delbet

"As long as chance rules," Arthur Koestler has written, "God is an anachronism."[1] Koestler's dictum is a sound conclusion . . . to a point. It is true that if chance *rules,* God cannot. We can go further than Koestler. It is not necessary for chance to rule in order to supplant God. Indeed chance requires little authority at all if it is to depose God; all it needs to do the job is to exist. The mere existence of chance is enough to rip God from his cosmic throne. Chance does not need to rule; it does not need to be sovereign. If it exists as a mere impotent, humble servant, it leaves God not only out of date, but out of a job.

If chance exists in its frailest possible form, God is finished. Nay, he could not be finished because that would assume he once was. To finish something implies that it at best was once active or existing. If chance exists in any size, shape, or form, God cannot exist. The two are mutually exclusive.

If chance existed, it would destroy God's sovereignty. If God is not sovereign, he is not God. If he is not God, he simply *is* not. If chance is, God is not. If God is, chance is not. The two cannot coexist by reason of the impossibility of the contrary.

This book, however, is not about God. It is about chance. It is about the existence of chance and the nature of chance.

What Is Chance?

We begin by asking the simple but critically important question, What is chance? Because this question is so critical, however, I think it important first to explain why the definition of *chance* is so crucial.

Words are capable of more than one meaning in their usage. Such words are highly susceptible to the unconscious or unintentional commission of the fallacy of equivocation. Equivocation occurs when a word changes its meaning (usually subtly) in the course of an argument. We illustrate via the classic "cat with nine tails" argument.

Premise A. No cat has eight tails.
Premise B. One cat has one more tail than no cat.
Conclusion: One cat has nine tails.

We see in this "syllogism" that the word *cat* subtly changes its meaning. In Premise A "no cat" signifies a negation about cats. It is a universal negative. In Premise B "no cat" is suddenly given a positive status as if it represented a group of comparative realities. Premise B assumes already that cats have one tail per cat. If we had two boxes, with one box empty and the second containing a single cat, we would expect to find one more cat in that box than in the empty one. If cats normally have one tail, we would expect one more cat's tail in one box than in the other.

The conclusion of this syllogism rests on the shift from negative to positive in the phrase *no cat*. The conclusion rests upon equivocation in the first premise. "No cat" is understood to mean a class of cats (positively) that actually possesses eight tails.

Such equivocation frequently occurs with the use of the word *chance*. We find this in the writings of philosophers, theologians, scientists—indeed pervasively. Here's how it works.

On the one hand the word *chance* refers to mathematical possibilities. Here *chance* is merely a formal word with no material content. It is a pure abstraction. For example, if we calculate the odds of a coin-flip, we speak of the chances of the coin's being turned up heads or tails. Given that the coin doesn't stand on its edge, what are the chances that it will turn up heads or tails? The answer, of course, is 100%. There are only two options: heads and tails. It is 100% certain that one of the two will prevail. This is a bona fide either/or situation, with no *tertium quid* possible.

If we state the question in a different manner, we get different odds or chances. If we ask, "What are the chances that the coin will turn up heads?" then our answer will be "Fifty-fifty."

Suppose we complicate the matter by including a series of circumstances and ask, "What are the odds that the coin will turn up heads ten times in a row?" The mathematicians and odds-makers can figure that out. In the unlikely event that the coin turns up heads nine consecutive times, what are the odds that it will turn up heads the tenth time? In terms of the series, I don't know. In terms of the single event, however, the odds are still fifty-fifty.

Our next question is crucial. How much influence or effect does chance have on the coin's turning up heads? My answer is categorically, "None whatsoever." I say that emphatically because there is no possibility, real or imagined, that chance can have any influence on the outcome of the coin-toss.

Why not? Because chance has no power to do anything. It is cosmically, totally, consummately impotent. Again, I must justify my dogmatism on this point. I say that chance has no power to do anything because it simply is not anything. It has no power because it has no being.

I've just ventured into the realm of ontology, into metaphysics, if you please. Chance is not an entity. It is not a thing that has power to affect other things. It is no thing. To be more precise, it is *nothing*. Nothing cannot do something. Nothing is not. It has no "isness." Chance has no isness. I was technically incorrect even to say that chance is nothing. Better to say that chance is not.

What are the chances that chance can do anything? Not a chance. It has no more chance to do something than nothing has to do something.

It is precisely at this point that equivocation creeps (or rushes) into the use of the word *chance*. The shift from a formal probability concept to a real force is usually slipped in by the addition of another seemingly harmless word, *by*. When we say things happen "by chance," the term *by* can be heard as a dative of means. Suddenly chance is given instrumental power. It is the *means* by which things come to pass. This "means" now assumes a certain power to effect change. Something that in reality is nothing now has the ability or power to do something.

Stanley L. Jaki in *God and the Cosmologists* gives a vigorous critique of the sloppiness by which the word *chance* is used in modern scientific and philosophical discussion. His chapter entitled "Loaded Dice" is a *tour de force* on the subject. In canvassing current cosmo-

Stanley L. Jaki

logical science, Jaki provides some astounding quotes. What follows is an analysis of some of these citations.

Delbet's Soft Pillow

Jaki cites Pierre Delbet's work *La science et la réalité,* published in 1913: "Chance appears today as a law, the most general of all laws. It has become for me a soft pillow like the one which in Montaigne's words only ignorance and disinterest can provide, but this is a scientific pillow."[2]

Jaki calls this "the softest 'philosophical' pillow in all scientific history," adding that "once more, as so often in that history, most successful mathematical formulas served as magic tools for making shabby philosophizing a most respectable attitude."[3]

Jaki's choice of descriptive terms is apropos. He speaks of "magic tools." The customary tool of the magician is the magic wand. The wand is waved over the empty hat accompanied by such incantations as "Abracadabra" and "Voilà!" The rabbit appears—*ex nihilo*. With this feat of prestidigitation the magician violates the oldest and most inviolate law of science: *Ex nihilo nihil fit* ("Out of nothing, nothing comes").

The magic, however, resides neither in the wand nor in the incantation. It is done via illusion. The trick rests upon the power of a mirror. The magician's hat (or box) is neatly divided into two compartments separated by a mirror. The cover or lid is opened halfway for the audience to inspect. Even if one peers into the hat from close range, he seems to behold a completely empty hat. What he actually views is half an empty hat and the mirror image of the empty half, which looks like a whole empty hat. Concealed behind the mirror is half a hat full of a rabbit (probably uncomfortable enough to agitate animal-rights devotees).

The trick requires only a modicum of sleight-of-hand dexterity to pull off. It is accomplished easily once one has the proper "magic tools." When scientists attribute instrumental power to chance, they have left the domain of physics and resorted to magic. Chance is their magic wand to make not only rabbits but entire universes appear out of nothing.

Delbet's metaphor is an apt one. Chance is indeed a "soft pillow." Pillows are used for sleeping as an aid to comfort. The soft pillow of chance has introduced a whole new era of dogmatic slumber. When the scientist dreams of chance he is dreaming of nothing, which, as Martin Luther once declared, "is not a little something."

Puritan theologian and philosopher Jonathan Edwards once mused, "Nothing is the same that the sleeping rocks dream of."[4] Now Delbet offers a comfortable pillow to aid the rocks in their reverie. However, even in our wildest dreams chance, which is nothing, cannot do something. To attribute causal power or any power to chance is to suffer from rocks in one's head. The pillow image suggests a lapse into unconsciousness by which otherwise brilliant thinkers take a nap. Their rational faculties have gone to sleep, blissfully imitating the rocks while they dream of nothing.

Though I poke fun at Delbet's soft pillow, it is ultimately no laughing matter. I don't think Jaki is indulging in hyperbole when he calls this concept of chance "the softest 'philosophical' pillow in all scientific history."[5] The soft pillow has not induced a sweet dream; it has provoked a nightmare.

The nightmare is not so much one about theology or philosophy, though it touches heavily on both; it is a nightmare for natural science. It reduces scientific investigation not only to chaos but to sheer absurdity. Half of the scientific method is left impaled on the horns of chance. The classical scientific method consists of the marriage of induction and deduction, of the empirical and the rational. Attributing instrumental causal power to chance vitiates deduction and the rational. It is manifest irrationality, which is not only bad philosophy but horrible science as well.

Perhaps the attributing of instrumental power to chance is the most serious error made in modern science and cosmology. It is certainly the most glaring one. It is serious because it is a patently false assumption that, if left unchallenged and uncorrected, will lead science into nonsense.

When Immanuel Kant read the works of David Hume, he exclaimed that he was awakened from his dogmatic slumbers. His *Critique of Pure Reason* was written with the purpose of saving empirical science from Hume's thoroughgoing skepticism. We need a similar awakening among cosmologists today who will

Immanuel Kant
(1724–1804)

lift their nodding heads from Delbet's soft, comfortable, but deadly pillow.

The Problem with Self-Creation

When the magic wand of chance is waved often enough and the pillow is soft enough, the second law that is transgressed is the law of noncontradiction. Magic and logic are not compatible bed fellows. Once something is thought to come from nothing, something has to give. What gives is logic.

To argue that something comes from nothing requires the denial of the law of noncontradiction. The law states simply that A cannot be A and non-A (-A) at the same time and in the same relationship. Something can be A and B at the same time but not in the same relationship. I can be a father (A) and a son (B) at the same time, but not in the same relationship.

For something to come from nothing it must, in effect, create itself. Self-creation is a logical and rational impossibility. For something to create itself it must be able to transcend Hamlet's dilemma, "To be, or not to be." Hamlet's question assumed sound science. He understood that something (himself) could not both be and not be at the same time and in the same relationship.

For something to create itself, it must have the ability to be and not be at the same time and in the same relationship. For something to create itself it must be *before* it is. This is impossible. It is impossible for solids, liquids, and gasses. It is impossible for atoms and subatomic particles. It is impossible for light and heat. It is impossible for God. Nothing anywhere, anytime, can create itself.

A being can be self-existent without violating logic, but it cannot be self-created. Let's summarize the train of thought we are following. The assertions I am making include the following:

1. Chance is not an entity.
2. Nonentities have no power because they have no being.

3. To say that something happens or is caused by *chance* is to suggest attributing instrumental power to nothing.
4. Something caused by nothing is in effect self-created.
5. The concept of self-creation is irrational and violates the law of noncontradiction.
6. To persist in theories of self-creation one must reject logic and rationality.

I grant that bold claims to self-creation are somewhat rare in scientific discussion. Usually the concept of self-creation is elliptical or camouflaged by obfuscatory language. The rose of self-creation usually blossoms under another name. The language of studied ambiguity triumphs here.

The French Encyclopedists solved the problem of language by masking the concept of self-creation under the rubric of "spontaneous generation." The term may be legitimately used to refer to sudden generation via imperceptible causes. It frequently functioned in the past, however, as a cloak for the concept of self-creation. Spontaneous generation for Denis Diderot made the God-hypothesis unnecessary. Indeed he was right. If a universe can spontaneously generate itself, who needs a Creator?

By the time I went to grammar school, the idea of pure spontaneous generation (à la self-creation) was largely discredited. Our science teacher smiled benignly at the folly of previous generations (unspontaneously generated generations) who had entertained and propagated such nonsense. The obituary for spontaneous generation, however, was premature. The naked concept continues to strut about like the unclad emperor of yore.

Recently I read a statement from a Nobel laureate in physics who declared that the days of speaking of spontaneous generation are over. He urged his readers to abandon the notion. He said henceforth we must speak of *gradual* spontaneous generation.

I am not exactly sure what gradual spontaneous generation means. Does it mean that something cannot create itself quickly? Is all that is lacking sufficient time to accomplish the task? Perhaps it is too much to expect from nothing that it generate something suddenly. But given enough time, it can do the job.

Wald's Miracles

Jaki cites the Nobel-laureate George Wald: "One has only to wait: time itself performs the miracles." "Given so much time," continued Wald, "the 'impossible' becomes possible, the possible probable, and the probable virtually certain."[6] Here is magic with a vengeance. Not only does the impossible become possible; it reaches the acme of certainty—with time serving as the Grand Master Magician.

In a world where a miracle-working God is deemed an anachronism, he is replaced by an even greater miracle-worker: time or chance. I say these twin miracle-workers are greater than God because they produce the same result with so much less, indeed infinitely less, to work with.

God is conceived as a self-existent, eternal being who possesses intrinsically the power of being. Such power is a sufficient cause for creation. Time and chance have no being, and consequently no power. Yet they are able to be so effective as to render God an anachronism. At least with God we have a potential miracle-worker. With chance we have nothing with which to work the miracle. Chance offers us a rabbit without a hat and—what's even more astonishing— without a magician.

That the concept of self-creation persists almost unchallenged can be illustrated by a news report I heard on the radio when the Hubble spacecraft was launched. The report quoted a noted scientist who declared, "Fifteen to seventeen billion years ago the universe exploded into being." The operative words here are the last three, "exploded into being." This is an assertion loaded with ontology. It is one thing to say that billions of years ago the universe experienced a massive explosion by which its structure and shape underwent massive changes. It is quite another thing to assert that it exploded *into being*. When something goes into something, it is moving from somewhere else. When I walk into my house, I am moving *out* of something else. Whence does something move when it moves into *being?* The only logical alternative is non-being. Does the statement mean that fifteen billion years ago the universe exploded from nonbeing into being? That's certainly what the statement implies. If so we can hardly resist the inference that that which exploded, since it was not yet in being, was nonbeing, or nothing. This we call self-creation by another name.

This is so absurd that, upon reflection, it seems to be downright silly. It is so evidently contradictory and illogical that it must represent a straw-man argument. No sober scientist would really go so far as to suggest such a self-contradictory theory, would they?

Unfortunately they would and they do. This raises questions about the soberness of the scientists involved. But generally these are not silly people who make such silly statements. Far from it. They number some of the most well-credentialed and erudite scholars in the world, who make a prophet out of Aristotle when he said that in the minds of the brightest men often resides the corner of a fool. In other words, brilliant people are capable of making the most foolish errors. That is understandable, given our frailties as mortals. What is not so understandable are the ardent attempts people make to justify such foolishness. The worst such attempt at justification is to justify nonsense by assailing reason itself. They attempt to give a reason for their irrationality.

Bohr's Great Truth

Someone as noted as Niels Bohr took this route with great gusto. Bohr's famous dictum is "A great truth is a truth of which the contrary is also a truth." So confident was Bohr of this statement that he emblazoned his coat of arms with the Latin motto *Contraria sunt complementaria.*[7] Bohr once argued that the two statements "There is a God" and "There is no God" are equally insightful propositions.

Perhaps the two propositions are equally "insightful," but they cannot be equally true. Indeed they could only equal each other in insight value if we deem false insights equal in value to true insights.

What is dismissed here along with the law of noncontradiction is perhaps the strongest formal argument in logic, the argument of the impossibility of the contrary. The impossibility of the contrary is basically a simple restatement of the law of noncontradiction. The impossibility of the contrary means that if A is, non-A cannot also be at the same time and in the same relationship.

Bohr's dictum was not so much an act of arrogance as it was an act of desperation to justify the unjustifiable. When scholars deny the law of noncontradiction, they do it selectively. That is, they do it when it suits them, when it is necessary to escape a logical trap. When logic snares us, the temptation is to retreat into denial. We deny the ensnarement by denying the trap that snagged us.

One final observation of the concept of self-creation. It is a concept that is analytically false. An analytically false statement is false by definition. To define a husband as an unmarried man or a triangle as a four-sided figure is to commit analytical falsehood. Analytically false statements are adjudged to be false not only because they are unintelligible, but because they are nonsense statements. They are not nonsense because they are unintelligible; they are unintelligible because they are nonsense. Empirical scientists may disparage philosophy, ontology, and epistemology, but they cannot escape them. Science involves the quest for knowledge. Any such quest, by necessity, involves some commitment to epistemology. The epistemology of irrationalism is fatal to all science because it makes knowledge of anything impossible. If a truth's contrary can also be true, no truth about anything can possibly be known.

The Mask
of Ignorance

"Let us stop talking of chance or luck, or at most speak of them as mere words that cover our ignorance."

Jacques Bossuet

The term *ignorant* is sometimes used pejoratively to describe someone who is crude or crass. Its primary meaning, however, has to do with the question of knowledge, or the lack of it. The root comes from the Greek word *gnōsis*, which means "knowledge." To be "agnostic" in the Greek sense is to be *a-gnōsis*, "without knowledge." The Latin equivalent is *ignoramus*.

Stanley L. Jaki begins his chapter "Loaded Dice" with a quotation from Paul Janet: ". . . chance is a word void of sense, invented by our ignorance."[1] I quarrel with Janet's statement about chance. Certain aspects of it are true, but one is in error. It is true that *chance* is a word. It is also true that it is invented by (or because of) our ignorance. It is not true, however, that the word is void of meaning.

Chance is a perfectly useful and meaningful word. We have already seen its valid use as a concept referring to mathematical possibilities. Its use in ordinary language is meaningful. When we say that two people met by chance in a railroad station, we describe a meaningful situation that we often use the word *coincidence* to describe. Recently I arrived by train at Chicago's Union Station. My train from Washington,

D.C., arrived about the same time commuter trains from the Chicago suburbs were discharging their passengers. In this mass of humanity I bumped into an old friend named Al. This serendipitous encounter was unplanned by either of us. We had no scheduled rendezvous. It was a "chance encounter." The coincidence was compounded later in the day. I returned to Union Station to catch my connecting train to Los Angeles. As I walked to the boarding area, I bumped into Al again. He was on his way home.

Here the term *chance* describes an event that was not planned by the parties involved. However, it was not an event without a cause. There were many causal factors involved that had us both at that place at that time. But the crucial point is that neither of us was there by the causal power of chance.

If we return to the illustration of the coin-toss, we remember that *chance* describes the mathematical possibilities of the outcome. Here "chance" serves as a kind of utilitarian shorthand. It is a kind of Ockham-like razor to simplify complex matters. When the referee stands at midfield before a football game with a coin poised on his thumb and asks the respective captains to "call it," the captains are not about to witness a controlled scientific experiment. They do not know if the coin to be tossed is face-up or face-down. They lack the knowledge of several other factors: the density of the atmosphere, the amount of force that will be exerted by the referee's thumb, the number of revolutions the coin will make, where the referee will catch the coin, whether the coin will be turned over after it is caught, or even if the coin will fall to the ground uncaught. With all these variables to consider, the captains are ignorant of the outcome in advance. Yet they are required to choose, to "predict" the outcome. They take their chances and make a call. Their decision will influence the game as the kicking and receiving teams will be determined.

Chance itself has no influence on the outcome of either the coin-toss or the subsequent game. The decisions made, reckoned by the players, will influence the game. They take their chances and make their call.

If we could conduct a closed experiment with a coin-toss in a vacuum by a fixed point of beginning from a fixed position, with a fixed number of revolutions and a fixed method of completion, and if the controlled experiment were repeated, the fifty-fifty odds would change dramatically. Indeed if the experiment were controlled enough, to wager against a repeated result given fifty-fifty odds would be the nadir of foolishness. It would be like betting against an instant replay of a horse race on the grounds that the odds are against the long-shot winner of the race duplicating his feat in the replay.

The term *chance* is also useful to describe games called "games of chance." Card games, dice games, and the like involve uncertain outcomes that are measured by probability quotients translated into odds. Games of chance, however, are not without their element of skill. The skilled card player pays close attention to the cards being played. (This is why card-counters are banned from certain gaming tables in Las

Vegas.) The more observant card player is able to reduce the ignorance factor involved and enhance his "chances" of winning. Likewise the skilled dice player masters the odds and plays accordingly. The skilled bridge player knows the odds of card splits that stand against him. In a particular move if he plays the percentages he may lose. In the long run, however, by playing the percentages he will win more than he will lose. In other words, his knowledge will make his guesses "luckier" in the long run.

But in all this, chance, though a meaningful term, is not a causal factor in any of the above transactions. It has no more power to influence the roll of the dice or the toss of a coin than it does the creation of a universe. When Janet says that chance is "invented by our ignorance," he is closer to the mark. Aristotle allowed for the formal significance of chance but not its material significance. Jaki comments: ". . . when it comes to material causality, Aristotle allows no chance and quite logically. Chance as a material cause would imply for him the rise of something out of no antecedent material cause, that is, out of nothing."[2]

Likewise Thomas Aquinas rejected chance as a material causality because it would imply that matter can arise spontaneously—that is, by chance or without cause—out of nothing.[3]

Jaki cites several allusions to the relationship between chance and ignorance. The list includes several notable figures.

From *The Crucifixion* by Fra Angelico, in the Refectory, S. Marco, France

Thomas
Aquinas
(1225–1274)

Bossuet's "Mere Word"

"Let us stop talking of chance or luck," wrote Jacques Bossuet, "or at most speak of them as mere words that cover our ignorance."[4] Here Bossuet describes chance as a "mere word." To say a word is "mere" indicates that the word is a name for something that is not a thing. Being a "mere word" means it has no ontological status. It is *nomina*, not *res*. (It is not by chance that the current debate in cosmology often enters the arena of the classical debate between nominalism and realism.) The fatal flaw of chance scientists is that they impute real being to the name *chance*.

When chance is said to cover our "ignorance," it does not imply a negative judgment on those who suffer from such ignorance. There is no shame in not knowing how the coin will turn up in an ordinary, uncontrolled coin-toss. The ignorance of the outcome is covered, thereby cutting the Gordian knot.

Hume's "Ignorance of Real Causes"

"Chance," according to Jaki's paraphrase of David Hume, "is only our ignorance of real causes."[5] Such a citation from Hume, who is frequently thought to be the demolisher of causality, may be surprising. We will return to Hume's treatise on causality in much greater detail later. For now let us be content to observe that Hume adds something specific to the notion of chance's covering ignorance. He spells out what is only tacit in Janet's and Bossuet's linking of chance with ignorance. He names the baby, unwilling to allow it to continue as a bastard child. The specific ignorance that the word *chance* covers is, as Hume asserts, the ignorance of *real* causes. Here chance is linked to two critical factors: causality and reality.

Throughout this book I have challenged the imputation of instrumental causal power to chance. To say things happen by chance is to make a statement about reality. It is a false statement about reality, a statement made in ignorance. Hume speaks of real causes. Presumably a real cause differs from an unreal cause or imagined cause. When we attribute a false cause or no cause to an event, we exhibit ignorance.

The only quarrel I have with Hume's statement is his use of the term *only*. He says chance is "only our ignorance." Insofar as the statement says something about chance, it is perfectly cogent and accurate. "Chance is only ignorance" means that that's all it is. It is not an entity with causal power. Here "only" functions like Bossuet's "mere." The ellipsis when spelled out translates: "Chance is only a word for our ignorance."

On the other hand if we read the statement with a different accent, it can be construed to mean, "Chance is a matter only of our ignorance." Since I doubt if this is what Hume meant, my quarrel dissolves. I labor this point only to give one caveat. The term *only* if applied to ignorance may suggest a trivial matter of ignorance. It may be construed to say, "Chance is only our ignorance of real causes, and our ignorance of real causes is an insignificant matter." But if our ignorance of real causes prompts us to substitute chance for real causes, then the matter becomes one of monumental significance. Chance as a substitute for real causes not only obscures real causes; it threatens the very heart of scientific inquiry, because ignorance expands to such a degree that it ends up in attempts to justify irrationality.

Voltaire's "Unknown Cause"

"What we call chance," wrote Voltaire, "can only be the unknown cause of a known effect."[6] There are some intriguing aspects to this observation. One aspect is that it is written by a man closely identified with the period of the Enlightenment. A vital legacy from that era was emphasis on the analytical method of science. This method may be defined as a quest for "the logic of the facts." The Enlightenment methodology ushered in a major breakthrough in empirical, inductive science. The search comprised basically two parts: first, the compilation of data or facts discovered and verified by observation, experimentation, measurement, and the like; and second, the analysis of the logic of these facts. The question was one of integration. Enlightenment science assumed a certain coherence or order to reality. It sought paradigms that would explain or make sense of individuated bits or facts. If certain facts could not fit the prevailing model, new models were sought and constructed to account for them. This was the modern version of the ancients' desire, à la Plato, to "save the phenomena."

In every paradigm ever constructed, including present ones, certain pesky anomalies are present. Ours is not the first scientific generation to produce an anomaly-free paradigm. Anomalies don't fit the logic of the paradigm. When they arise, we have three options: (1) deny the facts, (2) deny logic, or (3) modify the paradigm. What is deadly is an attempt to achieve 3 by using 1 or 2 (or both).

Anomalies represent present mysteries. They are unsolved problems. An easy solution to mystery is to give it another name: *chance*. Voltaire saw *chance* as a word-substitute for the unknown, again a cover-up for ignorance. The problem is one that confuses mystery and contradiction. All contradictions are mysterious. Not all mysteries are contradictions. To say that the cause of a known effect is unknown is to say that the cause remains a mystery. To say that the cause of a known effect is chance is to say that the cause is a contradiction. It is to say that the effect has no cause, which is a contradiction in terms.

The terms *mystery* and *contradiction* are often confused even by the most lucid thinkers. They are easy to confuse because they have so many similarities.

All contradictions are mysterious in the sense that we cannot understand them. We cannot understand

Voltaire
(1694–1778)

them because they are inherently, intrinsically, and eternally unintelligible. Even God cannot understand a contradiction. That which is illogical is unintelligible.

Logic has no content. Aristotle did not create logic; he defined it. He did not treat logic as a self-contained science. Rather he called it the *organon* or "instrument" of all science. He saw logic as the necessary condition for intelligible discourse. It is a prescientific necessity for doing any science at all.

Countless times in countless places by countless people, Aristotle's *organon* has been denied. People say they believe in contradictions, but they simply do not know what they are talking about. I say they *do not know* what they are talking about because they *cannot know* what they are talking about. People can easily assert contradictions. But no one can understand a contradiction. They remain forever mysterious.

On the other hand a bona fide mystery is a truth that may be intelligible. Many past mysteries have been resolved by the discovery of new information. The great accomplishment of scientific inquiry is that so many former mysteries have now been resolved, and resolved in an intelligible manner. We hope for more such resolutions as our knowledge bank expands. We have no such hope for the future resolution of contradictions. If we encounter two contradictory statements, we know that one of them may be true and the other false. It is also possible that both are false. What is not intelligible is that both can be true.

When we encounter bona fide contradictions in our scientific inquiry, we can take comfort that at least one of them is erroneous. This arms us with the weapon we need to achieve new breakthroughs by analyzing the assumptions to discern the error.

There is no greater erroneous assumption muddying the waters of contemporary science than the assumption that chance has instrumental, causal power. Here contradiction runs wild under the seemingly harmless cloak of mystery.

When Voltaire said that chance "can only be the unknown cause of a known effect," he was saying something about what is possible. If chance can only be one thing (the unknown cause), it cannot be something else. Voltaire did not mean that chance is, in fact, a cause. He said, "What we *call* chance. . . ." Voltaire, in effect, ruled out chance as an actual cause. He is saying that we use the word *chance* as a substitute word for a real cause that is yet unknown to us. But by calling the unknown cause "chance" for so long, people begin to forget that a substitution was made. It has been said that if we tell a lie often enough and boldly enough, people will begin to believe it. The assumption that "chance equals an unknown cause" has come to mean for many that "chance equals cause."

Voltaire's axiom was echoed by Claude Helvétius, who said, "I understand by chance the unknown chain of causes capable of producing such and such effect."[7]

Huxley's Incredulous Question

T. H. Huxley wrote, "Do they believe that anything in this universe happens without reason or without a cause?"[8] Huxley's question was made in response to those who tried to milk Darwinism for a wholesale endorsement of chance. Huxley was incredulous at such a prospect. Indeed to grant causal power to chance is not a matter of faith, but of credulity. What is implicit in the earlier quotations becomes explicit in Huxley, namely the link between causality and reason. This link will be further explored later. For now we merely note in passing that Huxley made the association.

In a sense Huxley was trying to defend Charles Darwin from Darwinists. Darwin himself once wrote to J. D. Hooker, ". . . I cannot look at the universe as the result of blind chance. . . ."[9]

Chance, of course, is in worse condition than being blind. Chance is deaf, dumb, and impotent as well.

Peirce's "Contradiction"

Charles S. Peirce wrote that "a contradiction is involved in the very idea of a chance-world."[10] In his most pragmatic moments, Peirce could not abide contradiction. He took no chances with chance.

The above citations raise an important question. Why, if the notion of chance as a real force is ruled out absolutely by such learned men of the past, does it persist in contemporary cosmological thinking? For

Photograph by Maull & Fox ca. 1854

Charles Darwin
(1809–1882)

the most part the scholars cited were not men carry-
ing a brief for theology. Most may be identified as
philosophers and thereby can be ignored in such an
antimetaphysical and anti-intellectual age as ours.
The ease with which logic is abandoned or insulted
reflects in part the modern rupture between science
and philosophy.

Modern science, in many quarters, has taken delib-
erate flight from philosophy. Scientists' motives are
mixed. Some do it out of frustration with the nonem-
pirical nature of much philosophical speculation. Others do it to escape accountability to the bar of reason.
Philosophers are the watchdogs of reason (at least they

were until they joined the flight by embracing existential thought).

There is, however, a more compelling reason for this flight into the absurd. It is the impact of quantum mechanics and Werner Heisenberg's uncertainty principle on modern scientific research.

A Quantum Leap

"Only by reconciling the two seemingly irreconcilable areas of physics can theorists hope to find a unified field theory that will explain the workings of the entire universe."

•

John Boslough

n his fascinating volume *Coming of Age in the Milky Way*, Timothy Ferris makes this observation regarding quantum indeterminacy: "Quantum indeterminacy may have nothing to do with human will, but as a matter of philosophical taste there are good reasons to celebrate the return of chance to the fundamental affairs of the world."[1] We notice that in this assertion Ferris makes no mention of good reasons to return chance to the world. (He states his reasons elsewhere.) Here he speaks of good reasons for *celebrating* the return of chance.

This statement intrigues me inasmuch as I am mourning rather than celebrating the return of chance. I am further intrigued by virtue of someone's celebrating the return of nothing, as we have already seen that chance is nothing. The celebration is indeed much ado about nothing.

Why the celebration? Ferris indicates that his celebration is a matter of philosophical taste, a highly subjective matter for which he seeks to give objective reasons. He qualifies his statement by adding: "Of course, the test of a scientific theory has to do less with whether one finds it philosophically palatable than with whether it works. Quantum physics works."[2]

Celebrating Success

The workability of quantum physics is not Ferris's only reason for celebrating the return of chance, but it is clearly his chief reason.

We ask simply, "It works for what?" Bringing chance back into the affairs of the world may work in terms of providing useful models of predictability of atomic behavior and the like. Chance as a calculation of probability factors certainly "works" in a bridge game or dice bet, which we have seen.

As an aid to mathematical models, chance surely works. As an aid to grasping real states of affairs, it fails—and fails miserably. Pragmatism may be well served by attributing causal power to chance, but truth and subsequently science are negotiated away in the process.

"Unreal" mathematical models have worked in the past, but usually to the detriment of the progress of science. Exhibit A would be the complex concept of intermeshing crystalline spheres that was at the heart of the Ptolemaic cosmology (see fig. 1). The fictional spheres "worked" to save the phenomena as they were perceived by the ancients. Even Ferris remarks:

> The system was ungainly—it had lost nearly all the symmetry that had commended celestial spheres to the aesthetics of Aristotle—but it worked, more or less. Wheeling and whirring in Rube Goldberg fashion, the Ptolemaic universe could be tuned to predict almost any observed planetary motion—and when it failed, Ptolemy fudged the data to make it fit. . . .

The price Ptolemy's followers paid for such precision as his model acquired was to forsake the claim that it represented physical reality. The Ptolemaic system came to be regarded, not as a mechanical model of the universe, but as a useful mathematical fiction.[3]

"Useful mathematical fiction": this phrase captures the spirit of the celebration of the grand return of chance. If chance returns as a useful mathematical fiction, then no harm is done. No harm, no foul.

We must be extremely careful, however, of the price tag of such fiction. We must also forsake the claim that chance represents physical reality. As a causal force, chance remains, ever and always, a fiction.

Ptolemy
(A.D. 2nd century)

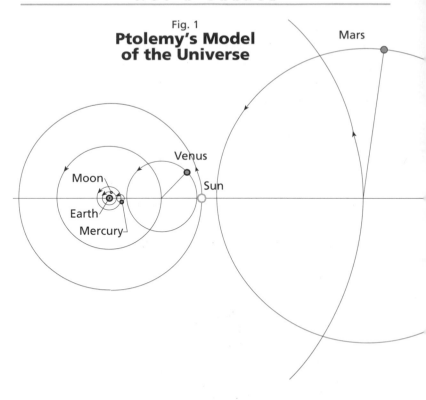

Fig. 1
Ptolemy's Model of the Universe

"The aim of the theory [of Ptolemy], then," Ferris adds, "was not to depict the actual machinery of the universe, but merely to 'save the appearances.' Much fun has been made of this outlook, and much of it at Ptolemy's expense, but science today has frequent recourse to intangible abstractions of its own. . . . It should be said in Ptolemy's defense that he at least had the courage to admit to the limitations of his theory."[4]

Those today who insist on assigning to chance real power rather than a fictional function lack either Ptolemy's understanding or his courage.

It is the workability of chance that excites Ferris. At

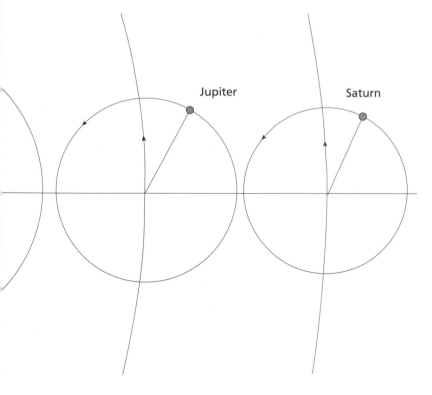

the Laurel Valley Country Club in Ligonier, Pennsylvania, a golf professional named John Rock labored
for many years. John had a strange quirk. When I
entered the pro shop, he never greeted me with such
standard salutations as "Hello" or "How are you?" John
always greeted me with a simple question: "Does it
work?" Knowing the routine, I always gave the same
reply, "Yes!" to which John responded, "Then use it!"

John Rock was a master putter. He told me there is
no single correct way to putt. He constantly repeated
the maxim, "If it works, use it!"

There is more than utility to modern quantum the-

ory. To set the issues before us, let us take a brief recon-
naissance of the historical scene.

Small Beginnings

Quantum theory traces its origins to the work of Max
Planck, who presented in 1900 the hypothesis that
energy comes in discrete units called "quanta."

From small beginnings quantum theory expanded
into virtually every area of physics, bringing in its wake
a revolution. The watershed moment came in 1927
with the postulation of the indeterminacy principle by
German physicist Werner Heisenberg:

> Heisenberg found that one can learn either the exact
> position of a given particle or its exact trajectory, *but
> not both*. . . . In classical atomic physics it had been
> assumed that one could, in principle, measure the pre-
> cise locations and trajectories of billions of particles—
> protons, say—and from the resulting data make exact
> predictions about where the protons would be at some
> time in the future. Heisenberg showed that this assump-
> tion was false—that we can *never* know everything
> about the behavior of even *one* particle, much less myr-
> iads of them, and, therefore, can never make predic-
> tions about the future that will be completely accurate
> in every detail. This marked a fundamental change in
> the world view of physics. It revealed that not only mat-
> ter and energy but knowledge itself is quantized.[5]

The bogyman in the above passage is the word *never*.
That we can *never* know the behavior of one particle

Photograph ca. 1925

Werner Heisenberg
(1901–1976)

is a strong assertion. It's not so much that it poses a limit to human knowledge—there is nothing new about that. It is the "why" of the limitation that has evoked so much heated controversy.

To get at the "why" we must probe further into quantum theory. Again Ferris summarizes:

> The more closely physicists examined the subatomic world, the larger indeterminacy loomed. When a photon strikes an atom, boosting an electron into a higher orbit, the electron moves from the lower to the upper orbit instantaneously, *without having traversed the intervening space.* The orbital radii themselves are quantized, and the electron simply ceases to exist at one point, simultaneously appearing at another. This is the famously confounding "quantum leap," and it is no mere philosophical poser; unless it is taken seriously, the behavior of atoms cannot be predicted accurately. . . .
>
> Those who find such considerations nonsensical are in good company. . . .[6]

I am enormously relieved by the final sentence in the above passage. I'd hate to think I'm all alone in judging the considerations avowed to be nonsensical. I only hope the good company increases in number. Do we have here an actual quantum leap of subatomic particles, or do we have a quantum leap in the thinking of scientists who are trying to understand the behavior of the particles?

I find nonsense in the above at two specific points. The first is in the assertion that electrons move into a higher orbit instantaneously, *without having traversed the intervening space.* This is a fascinating concept. Does it mean that the electrons move and don't move at the same time? Do they change positions without changing positions? Do they traverse space without traversing space? It would seem that to ask such questions is to answer them.

How is the conundrum solved? The electron does move magically, without jumping, tunneling, or end-running the intervening space. It gets from one place to another by simply "ceasing to exist at one point, while simultaneously appearing at another."

Now, we humbly ask, if the electron ceases to exist altogether, how do we know that it is the same electron that simultaneously appears somewhere else? Does the electron pass out of being and back into being? Is it destroyed and created all at the same time? Does it exist and not exist at the same time and in the same relationship? If so, science is finished, wrecked by maverick electrons that make knowledge of anything impossible.

If the electron goes out of existence and comes back into existence, what causes it to do so? Does it bring itself back into existence by the impossible act of self-creation? Does it come back into existence by the power of chance? If so, then we have something out of nothing, because even in Heisenberg's experiments chance is nothing.

Being There

But the passage doesn't specifically assert that the electron passes out of existence. It asserts that it ceases to exist *at one point.* That's true of any entity in motion.

If, as Martin Heidegger asserted, being (*Sein*) is always being there (*Dasein*), then there is always a certain "whereness" or location of finite entities. This finitude of location is assumed when it is asserted that the electron does not traverse the intervening space. Since it is not ever in the intervening space, its location is not infinite. Our maverick electron is not ubiquitous.

I often ask my students the simple question, "Where do you live?" If I ask the question in Orlando, I frequently get answers like "I live in Gainesville." I then ask, "Are you presently, at this moment, in Gainesville?" They reply, "No, of course not; I'm here in Orlando." Then I ask, "Are you, at this moment, alive?" With growing consternation they say, "Of course!"

My point is simple. We use language like "I live in Orlando" to refer to our place of residence. We are not always in our place of residence. Sometimes the lights are on but we are not home. Strictly speaking the only time I "live" in Orlando is when I am *in* Orlando. I live wherever I am at any given moment. When I leave Orlando I don't cease to exist. My being remains intact. I do cease to exist "in Orlando," since my "being" is no longer located there.

But a big difference remains between my moving from place to place and the apparent movement of Heisenberg's electrons. I cannot leave Orlando and *simultaneously* appear in Chicago. It takes me time to get from one place to another. Even Superman, who leaps tall buildings in a single bound, cannot make a quantum leap over a building. Nor can I move from Orlando to Chicago without traversing the intervening space.

The theory moves from the physical into the metaphysical. If the physical boundary of space is violated, then it must be a metaphysical phenomenon. To circumvent space in motion is a prodigious feat. To do it without the passage of time is even more prodigious.

If the quantum leap is literally *simultaneous,* then we have a bona fide contradiction. We have something that is here and not here at the same time and in the same relationship. If an electron or any other atomic particle can do that, it can do something even God can't do. God may be able to be ubiquitous, but even he cannot be somewhere and not be there at the same time and in the same relationship. If a theologian declared that God can be in Boston and not be in Boston at the same time and in the same relationship, he would be laughed to scorn by the scientific community. Please excuse this theologian for laughing at the attribution of this same ability to an electron.

A quantum leap is an illusion. Illusions may seem to be real; that's why they are called illusions. Yet we distinguish between a reality and an illusion. The illusion is judged to be illusion when what appears to be the case in fact cannot be the case.

No one is disputing the "appearance" of quantum behavior. Undoubtedly Heisenberg encountered a devilish problem of atomic-particle predictability. I can imagine his saying: "This is incredible. The electron seems to be disappearing from one orbit and appearing in another simultaneously and without traversing the intervening space. How in the world can I explain this apparent behavior?"

One thing is certain about the uncertain behavior of subatomic particles: at the moment I certainly cannot explain it. It remains a baffling mystery to me. I don't know how the electron gets from one orbit to another and does it so rapidly that it seems *virtually* simultaneous. I cannot say what causes this effect.

I am equally certain, however, that we must not attribute the cause of this mysterious activity to chance. To say that chance *does* it is to say that *nothing* does it. This is to say that nothing does something, which is nonsense. The concept of chance motion does not explain anything. It is a useless theory because it rests upon nothing.

Uncaused Effects

What is basically happening here is the tacit assertion that we can have effects without causes. I coauthored a book on apologetics[7] that was reviewed by a theological journal. The reviewer offered the following criticism: "The problem with Sproul is that he does not allow for an uncaused effect." I responded to this professor, skilled in philosophy and apologetics, by saying, "*Mea culpa*. I am guilty as charged. You are right, I do not allow for uncaused effects. Heretofore, however, I considered that an intellectual virtue and not a crime. I will be quick to repent of my stubborn refusal if you would just give me one single example of an uncaused effect." I am still waiting for his reply. To his credit he did not respond by citing quantum leaps of electrons as an example.

I have great respect for the philosopher who so criticized me. That he apparently allows for uncaused effects does not serve to open my mind to their possibility, but merely serves to remind me how easily even erudite scholars can become confused.

I do not allow for uncaused effects because uncaused effects represent a contradiction in terms. The idea of an "uncaused effect" is analytically false. It is a nonsense statement, akin to speaking of square circles and married bachelors. An "effect" is by definition something produced by an antecedent cause. If it has *no cause,* it is not an effect. If it is an *effect,* then it has a cause.

It is one thing to say that electrons behave in a certain way for uncertain reasons. It is another thing to say that they behave in a certain way for no reason. Again we can say with humility that they behave this way for no *apparent* reason or for some as-yet-*unknown* reason. We cannot say with humility that they behave this way for no reason or without cause.

To say that things happen for no reason or that effects take place without a cause is to speak with unmitigated and consummate arrogance. There is not a trace of humility in it. Why do I make such a severe allegation of arrogance? The allegation itself may appear arrogant.

I call it arrogance because it presupposes an attribute no mortal has, scientists or anyone else. It presupposes *omniscience.*

Omniscience is not required to say that there can be no effects without a cause. That is an assertion of formal logic. It requires not omniscience, but simply rationality. To say that an effect has no cause can easily be done by retreating into irrationality (and still retaining some humility). But to say that we know a given effect has no cause presupposes that we have full knowledge of every possible cause in the universe. That requires total knowledge of all that there is. If there is any gap in the knowledge, it remains possible that the unknown cause is in that gap.

This recapitulates the classic difference between the problem of positive verification of empirical assertions and the problem of their falsification. If we assert that "there is gold in Alaska," all that is required to verify the statement empirically is to find one piece of gold in Alaska, which theoretically could be achieved with one turn of the shovel.

Falsification is not so easy. If we say "There is no gold in Alaska," how do we falsify the negative? Of course it is falsified if its contrary is proven by the discovery of gold there. But suppose we can't find any gold there. How much of Alaska must be examined before we come to the certain conclusion that there is no gold there? Theoretically we could search every square inch of Alaska without finding gold. That would not prove that there is no gold there. We could have overlooked a minute piece of gold concealed in one of those square inches. To establish that there is *absolutely* no gold in Alaska requires that we have absolute knowledge of Alaska.

The problem of falsification is inherent in inductive science. No inductive process is ever absolutely complete. There always remains one more datum bit to consider. We generalize on the basis of known data and examples. If we discover one million black ravens and no white ravens, we generalize that "all ravens are black." But no one has exhaustively examined all ravens. Indeed we can never know that we have seen all ravens. After examining one million ravens, all of which are black, the next raven may be a white one.

William Poundstone writes:

> Generalizations are concealed negative hypotheses: "There is no such thing as a non-Y X"; or "The above rule has no exceptions." A generalization's contrapositive corresponds to the identical negative hypothesis.
>
> In an infinite universe, proving a negative hypothesis is a supertask. (If the universe is merely finite but very big, proving a negative hypothesis is a herculean labor so close to a supertask as to make no difference.) We are incapable of supertasks, and have reason to be suspicious of knowledge attainable only through supertasks anyway.[8]

To celebrate the return of chance into the affairs of this world (in the sense that chance causes quantum leaps and therefore "justifies" the negative hypothesis that nothing causes the behavior of subatomic particles) is an exercise in futility. Such justification via negative hypothesis is more than a supertask—it is an impossible task. It is a task that even Hercules could not perform. If Hercules were to return, we would not ask him to perform such a task. We would ask him to repeat one of the tasks he had already performed. We would ask him once again to clean the stables.

Celebrating Freedom

Again we ask why Ferris, for example, would be so sanguine about the return of chance. One answer is hinted at by John Boslough in his book *Stephen Hawking's Universe*:

> Quantum mechanics seems to suggest that the sub-atomic world—and even the world beyond the atom—has no independent structure at all until it is defined by the human intellect. (This view of the universe has similarities to Eastern philosophy, which has led, to the dismay of Hawking, to a wealth of popular literature such as Fritjof Capra's *The Tao of Physics* and Michael Talbot's *Mysticism and the New Physics* that attempts to link quantum physics with Eastern mysticism.) Physicists have been unable to reconcile this system with the view of the universe posited by general relativity. While general relativity allows for a perfect pointlike singularity at the beginning of time, quantum mechanics does not, for it prohibits defining at the same time the precise location, velocity, and size of any single particle or singularity. . . . Only by reconciling the two seemingly irreconcilable areas of physics can theorists hope to find the unified field theory that will explain the workings of the entire universe.[9]

The attraction of quantum theory is not limited to Eastern religion. It is alluring for any philosophical school that seeks escape from reason and the binders of the law of noncontradiction. It is a grand privilege to argue without the constraints of logic. To be free of causality is to be free of logic, and license is given for making nonsense statements with impunity. Dialectical theologians love it because now they can legitimately speak of one-handed clapping and declare that there is no God and Mary is his mother.

Now we are free to be inconsistent. We can give credence to the adage that consistency is the hobgoblin of small minds.

Boslough contends that *only* by reconciling the two *seemingly irreconcilable* areas of physics can we hope to find a unified field theory. This translates into saying this is the only way to have cosmos rather than chaos. An un-unified field is chaos.

Boslough sees such reconciliation as the *only* way to save science. If it is the only way, then it is a necessary condition. If Boslough is correct, then the scientific stakes are as high as they can be. Whatever benefits may accrue to Eastern mysticism or dialectical thought, the loss to science of a unified field is an enormous price to pay.

We notice that Boslough hopes for the reconciliation of two *seemingly* irreconcilable views. What if the views are not only *seemingly* irreconcilable but *actually* irreconcilable? Then there can be no reconciliation. The necessary condition cannot be met and we are left with chaos.

We are left with a dilemma akin to finding a universe that contained both an irresistible force and an immovable object. Is such a universe possible? Not if language means anything. The old poser of what happens if an irresistible force meets an immovable object is relevant. Let's explore the possibilities. If such an encounter took place, only two possibilities exist. Either the object would move or it wouldn't. There is no *tertium quid* possible. If the object moved, we would know immediately that we were incorrect in deeming it immovable. If it didn't move, we would know instantly that our "irresistible force" was in fact resistible and was misnamed.

We do not need an empirical experiment to prove that irresistible forces cannot coexist with immovable objects unless they are one and the same thing. We know this because the ideas are formal impossibilities as coexistents. The experiment could demonstrate *which* of our assumptions about reality is false, but no experiment is needed to know that at least one of the two does not exist.

If the two views of physics are actually irreconcilable, then we know that at least one of them is false and must be discarded. If they are to be reconciled, it won't be by appealing to nonsense views of chance.

It is the conflict between these two views that provoked the celebrated controversy between Niels Bohr and Albert Einstein.

The Voice
of Reason

"I, at any rate, am convinced that He *is
not playing at dice. . . ."*

•

Albert Einstein

iels Bohr's dictum "A great truth is a truth of which the contrary is also a truth" stirred up great controversy. The controversy still rages. Carl Sagan, in an appendix to his popular work *Cosmos,* writes: "To take a modern example, consider the aphorism by the great twentieth-century physicist, Niels Bohr: 'The opposite of every great idea is another great idea.' If the statement were true, its consequences might be at least a little perilous."[1]

Carl Sagan is not known for his timidity. But if ever he lacked temerity in speech, it is here with such a carefully guarded statement. He says, "If the statement *were* true . . ." The presence of "if . . . *were*" rather than "if . . . *was*" indicates a condition contrary to fact. By using *were* instead of *was* Sagan is not, I trust, merely making a stylistic distinction but is in fact considering Bohr's aphorism to be a condition contrary to fact.

Still the *if* persists, at least as a hypothetical possibility that Sagan is considering for the sake of argument. In this appendix he is discussing the ancient form of the *ad hominem* argument known as *reductio ad absurdum* (of which this present essay of mine is but a modern version).

If Bohr's statement were true, according to Sagan's timid caveat, its consequences *might* be *at least* a *little* perilous. *Might* be perilous? Sagan blinks here. There's no "might be" to it. It is not only possibly perilous but certainly perilous and far more than merely a *little* perilous. There is enormous, gigantic peril in such an aphorism. It represents a clear and present danger to all science and philosophy.

That Sagan is indulging in understatement, treading softly for the moment amid great peril, and that he clearly understands the real enormity of the peril is evidenced by his subsequent statements. Sagan warms to his task:

> For example, consider the opposite of the Golden Rule, or proscriptions against lying or "Thou shalt not kill." So let us consider whether Bohr's aphorism is itself a great idea. If so, then the converse statement, "The opposite of every great idea is not a great idea," must also be true. Then we have reached a *reductio ad absurdum.* If the converse statement is false, the aphorism need not detain us long, since it stands self-confessed as not a great idea.[2]

To take Sagan's musing to the next step, which is elliptical in his argument, we say that if the prohibition against homicide (which is the import of the biblical commandment "Thou shalt not kill") is a great truth, then an apodictic mandate for murder would be also a great truth. If "Thou shalt murder" is an equally great truth, then this truth would provoke more than a little peril to human life. But, because this is so self-

Niels Bohr
(1885–1962)

evidently the case, Sagan has succeeded in reaching the *absurdum* in the reduction of Bohr's aphorism.

Einstein's Response

Perhaps no one saw the peril in Bohr's thinking more clearly than Albert Einstein. His vociferous reaction cost him dearly in terms of personal reputation.

Nigel Calder, in his exposition of Einstein's life and thought *(Einstein's Universe),* considered Einstein's reaction an "appalling blind spot."[3]

Timothy Ferris relates Einstein's reaction to the Copenhagen understanding of quantum theory: "Einstein was deeply troubled by this aspect of the new physics. 'God does not play dice,' he said, and he argued that the indeterminacy principle, though useful in practice, does not represent the fundamental relationship between mind and nature."[4]

Ferris then includes a record of Einstein's correspondence with his colleague Max Born:

> I find the idea quite intolerable that an electron exposed to radiation should choose *of its own free will*, not only its moment to jump off, but also its direction. In that case, I would rather be a cobbler, or even an employee in a gaming-house, than a physicist.[5]

Einstein saw in the "intolerable idea" a reason for despairing of doing physics. When I argue that science is at stake in this controversy, I am comforted by the knowledge that no less a scientist then Einstein shared my concern.

Later Einstein told Born:

> Quantum mechanics is certainly imposing. But an inner voice tells me that it is not yet the real thing. The theory says a lot, but does not really bring us any closer to the secret of the "old one." I, at any rate, am convinced that *He* is not playing at dice. . . . I am quite convinced that someone will eventually come up with a theory whose objects, connected by laws, are not probabilities but considered facts. . . .[6]

I suspect that the "inner voice" nagging Einstein was the voice of his own reason. Calder and Bohr didn't

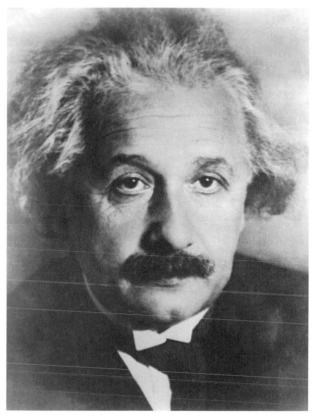

Albert
Einstein
(1879–1955)

think so. Bohr responded to Einstein's concept of God's
not throwing dice in an angry pique: "Stop telling God
what to do!" Bohr replied.[7]

Calder observes:

The Danish quantum theorist and his brilliant associates turned their backs on Einstein and showed that
God does indeed play dice: he has his gaming tables
in every atom and every cubic millimetre of empty
space. Flowering into the theories of anti-matter, of
nuclear physics, of electricity and the sub-atomic

forces, quantum mechanics became a luxuriant growth, more extensive and productive than the tidier gardens of relativity.[8]

Here Calder rescues God from the fate of being an anachronism. God is alive and well, spending his time playing dice. His glory is that he has no need to travel to Las Vegas for his sport. He "has gaming tables in every atom" and throughout "empty space."

For Calder the debate is settled. He has resolved the controversy by siding clearly with Bohr, who together with his brilliant associates "showed that God does indeed play dice."

Calder adds that "his revulsion from the quantum theory made Einstein's own work sterile." He concludes that Einstein "confused 'uncertainty' in its subatomic, statistical sense with 'uncertainty' about cause and effect, which he rightly abhorred."[9]

This final comment puzzles me. What does he mean that Einstein "rightly abhorred" uncertainty about cause and effect? Calder suggests that Einstein was right about the causal issue but wrong about the dice game. For Einstein the issue was one and the same. Calder can't have it both ways.

Stephen Hawking, at least at one point, sided with Bohr. He said, "God not only plays dice, but sometimes he throws them where they cannot be seen" (e.g., in black holes).

I once repeated this quote to Mortimer Adler, who snorted in reply, "He doesn't know what he's talking about."

Stanley L. Jaki observes about Einstein's dice metaphor:

> There is indeed some subtle misunderstanding of the nature of dice in Einstein's remark that God does not play dice. For in any throwing of dice it must be assumed that they cannot come to stop in mid-air or on their corners and edges, but only on their sides. . . .
>
> An implied acknowledgment of such limitations is Einstein's famous remark that the good Lord is subtle but not malicious. For the Lord's subtlety is revealed by the fact that nature is understandable and that it is natural for man to understand things. Of course, if man is taken for a chance product of blind evolutionary forces, the battle, be it waged by Einstein against the rule of chance, cannot be fought convincingly.[10]

In simple terms Jaki is asserting that the *rule* of chance cannot be defeated if it is allowed a single victory. If *anything* is really a *chance product,* then the battle is already over. Science falls in the battle. Here science suffers from intellectual hemophilia. Scratch science with chance and it bleeds to death. Again at the point of *ad nauseum* repetition, if chance can produce anything we can have something from nothing, which destroys both causality and logic with a single blow. We are left abandoned to ultimate, inexorable chaos.

Jaki continues:

> Those reluctant to draw the proper lesson about the contradictory character of randomness or chaos would

do well to look at themselves for a moment in the mirror of John von Neumann's words that strike at the very root of mathematically conceived randomness: "Anyone who considers arithmetical methods of producing random digits is, of course, in a state of sin."[11]

Von Neumann escalates the debate to the moral level. He sees a sin against reason as a sin against God. We may excuse the "sin" as a mere error of ignorance. Roman Catholic moral theology, however, makes a sharp distinction between vincible ignorance and invincible ignorance. Invincible ignorance is ignorance that cannot be conquered. It cannot be overcome. Such ignorance excuses and is not considered a sin. Vincible ignorance, however, is ignorance that can and should be overcome. It does not excuse. To attribute ontic power to chance is a matter of vincible ignorance. We should know better.

In describing recent scientific inquiries into the question of chaos, Jaki observes:

For insofar as those active in those studies "are looking for the whole," they renounce the notion of chaos. A chaos can never be a whole, that is a co-ordination of parts, without ceasing to be a chaos, properly so-called. Highly significant should seem the fact that none of those studies yielded a satisfactory definition of chaos. Circumlocutions about it, especially if loaded with esoteric mathematics, may impress even those who should know better. When a science reporter states about a major conference on chaos that it "had attained the status of buzz word, but few of the physicists attending the conference knew what it meant," he merely reveals his inability to pick up the truly valuable information. Rather he should have said that none of those physicists really knew what they were talking about. And on having heard them state that chaos is "an agent of order" or that there is a "sensitive chaos," he should have warned them about the dangers of linguistic confusion.[12]

Statements like "chaos is an agent of order" are examples of linguistic confusion. Such confusion is a serious problem. Indeed it may be *the* problem between classical physics and quantum physics. Linguistic problems will continue to plague us as anomalies continue to appear that do not fit existing paradigms. Until the paradigm is revised or expanded to accommodate the anomalies, we will tend to attempt to squeeze or force the anomalies into old categories, giving rise to linguistic confusion and even nonsense statements.

Here is where great caution is required plus at least a basic understanding of the function of language. When we probe the depths of language itself, we encounter mysteries that may baffle us almost as much as elements of quantum experiments.

Phenomenological Language

When an experiment is performed and results are calculated, we are still left with the problem of articulating adequately what we have perceived. One crucial word that is often used is *seem*. One use of language is that of describing things according to how they *appear* to us. We call this phenomenological language.

Phenomenological language has been a bugaboo in classical thought. The great debate between the Roman Catholic church and both Nicolaus Copernicus and Galileo Galilei might have been happily avoided if enough attention had been paid to phenomenological language. Is it false or erroneous to describe the sun as moving across the sky from east to west? It *appears* to the naked eye that the sun moves. Who would have imagined from simple observation that reality differs so greatly from our perception of it?

Even modern weather forecasters, whom we call meteorologists (with a nod to the outdated Aristotelian view of meteors), speak in terms of phenomenological language. They give us barometric pressure, wind velocity, precipitation probability quotients, and all sorts of modern technical jargon—then proceed to speak of "sunrise" and "sunset," as if we still held to a geocentric view of the universe.

Who perceives directly that the earth rotates on its axis? I don't feel like I'm on a planetary merry-go-round. The confirmation of the "reality" of the earth's rotation came about in rather serendipitous fashion.

When Copernicus speculated that the earth was moving, few people took him seriously. The idea seemed utterly incredible since it conflicted so powerfully with ordinary perceptions. It was a "quirk" in

Ferdinand Magellan's circumnavigation of the globe that provided the confirming evidence.

William Manchester in his book *A World Lit Only by Fire* tells the story: "The full significance of the great voyage was not grasped until much later, but its most

Brausewetter Pinxt

Nicolaus
Copernicus
(1473–1543)

profound implication had begun to emerge two months before the *Victoria*'s return to Spanish waters, when she was anchored off Santiago in the Cape Verde Islands."[13]

At this point in the voyage an apparently trivial argument broke out between the Portuguese and members of the shore party from Magellan's expedition. The two groups argued about which day of the week it was. According to the sailors it was Wednesday, July 9, 1522. But according to the calendar in Santiago it was Thursday, July 10.

The debate became furious. Both sides were sure they were correct, and both sides knew that it couldn't both be Wednesday and not be Wednesday at the same time and in the same relationship. Somebody was wrong. The people in Santiago were sure that their calendar makers had not made a mistake. The error had to have been committed by Magellan's crew.

Magellan's crew was equally certain they were not in error. Their data had been compiled not by one man but by two. Francisco Albo had kept a scrupulous ship's log every day during the voyage, and Don Antonio Pigafetta had kept a daily diary. Both of their datings agreed.

The keeping of such logs and diaries was not a casual matter. The purpose was not to be able to recount yarns for their grandchildren. There was a scientific motivation for the logs and diaries. Their writers were compiling critically important data for future expeditions. Theirs was an important kind of scientific documentation. Navigational charts, tides, positions, and depth-soundings were kept daily for future reference. Such logs were priceless in the business of exploration. Magellan himself was armed with data compiled by early explorers that had been likewise committed to logs.

Manchester writes: "Don Antonio was puzzled. It was inconceivable that he could have missed a day. He checked with Albo, who, on instructions from Magellan, had also kept a record of the days in his ship's log."[14]

The result of this dispute was that scientists of the sixteenth century toiled to reconcile the apparent contradiction. They unanimously agreed on the solution: "Copernicus, they concluded, was right. The earth was rolling eastward, completing a full cycle every day. Magellan and his men had been sailing westward, against that rotation; having traversed a full circle, the circumnavigators had gained exactly twenty-four hours."[15]

When couriers carried this information to the Vatican, the pope rejected the conclusion (as did many Protestant theologians):

Twenty-eight successive pontiffs agreed. It took the Church three hundred years to change its mind. Copernicus's *De revolutionibus* was removed from the Catholic Index in 1758, but the ban on Galileo's *Dialogue* continued until 1822, exactly three centuries after Albo's log and Don Antonio's diary had become available to the Holy See.[16]

The church could have avoided this black eye if it had allowed the Bible to speak in phenomenological language, which is a legitimate description of things as they appear.

With the advent of the microscope we know that things, in a certain sense, are not as they *appear* to us. The *appearances* are real. We are speaking too pessimistically to call them illusions. But appearances give us the "real phenomena," not necessarily the "real essences." With modern advances in technology our phenomenal horizons have broadened and deepened. Yet our perception still has limits. We are still searching for the "essence" of reality—for Immanuel Kant's "thing-in-itself."

Atomic theory reminds us that things are not always as they appear. Ferris observes:

> Our mental pictures are drawn from our visual perceptions of the world around us. But the world as perceived by the eye is itself exposed as an illusion when scrutinized on the microscopic scale. A bar of gold, though it looks solid, is composed almost entirely of empty space: The nucleus of each of its atoms is so small that if one atom were enlarged a million billion times, until its outer electron shell was as big as greater Los Angeles, its nucleus would still be only about the size of a compact car parked downtown. . . . The quantum revolution has been painful, but we can thank it for having delivered us from several of the illusions that afflicted the classical world view.[17]

What Ferris says here is helpful and insightful. His statements, however, need to be qualified a bit. He praises quantum physics for rescuing us from the illusions of the classical view. We need to be reminded that the classical view was not ignorant of the distinction between *phenomena* (outward appearances) and *essence* (reality). Aristotle himself made much of the difference between *substance* and *accidens*.

Aristotle, however, considered the *accidens* as real, not illusory. He retained a realistic link between essences and their outward appearances. So does Ferris. The solid appearance of the bar of gold is not a total illusion. The outward appearance of the gold bar is not caused by nothing. Its atomic structure is that of a gold bar, not an elephant.

We also ask what Ferris means when he says that the bar is "*composed* almost entirely of empty space." What is the essence of empty space? It appears to be nothing, but nothing has no composition.

Admittedly this is a quibble. I have no real objection to Ferris's manner of speaking. I am simply noting that even here he is speaking in phenomenological terms. His use of the term *illusion* is a bit shaky.

I encounter the "illusion" of solidness when I play golf. When stymied behind a tree, I invariably hear the unsolicited counsel of my partners: "Don't worry, trees are 90% air." What I can't figure out is why, when I hit the ball directly at the tree, it doesn't go through it safely 90% of the time.

Paradoxes and Contradictions

Perhaps it is because so much of our language is phenomenological that we encounter so many paradoxes. The term *paradox* is well suited for such language. In its etymological derivation the word comes from a Greek prefix and a root. The prefix *para-* means "alongside of," as in *para*medical and *para*legal. The root comes from the Greek *dokein*, "to seem, to think, or to appear." A paradox is something that "seems" or "appears" to be something else that is placed alongside of it. What is this "something else"? It is the *contradiction* or the *antinomy.*

In the fluid use of language, we sometimes encounter what Jaki calls "linguistic confusion." In current customary usage the terms *paradox, contradiction,* and *antinomy* often are interchangeable, functioning as synonyms. In classical terms, however, there is a crucial distinction among them.

Classically the words *contradiction* and *antinomy* are synonyms. Antinomy comes from the prefix *anti-* (against) and the Greek noun *nomos* (law). Etymologically an antinomy is something that is "against the law." What law? The law of noncontradiction. Antinomies are statements that violate the law of noncontradiction.

The term *contradiction* means virtually the same thing. It comes from the Latin prefix *contra-* (against) and *dicio* (to speak). Contradictions are statements that speak against each other and also against the law of noncontradiction.

Contradictions and antinomies involve nonsense statements that provoke no small degree of linguistic confusion. One philosopher described contradiction as "a charley horse between the ears."

On the other hand a paradox differs significantly from contradiction and antinomy. A paradox is something that *seems* like a contradiction but under closer scrutiny can be resolved. Herein lies the difference: paradoxes can be resolved, contradictions cannot.

It is one thing to say that reality is paradoxical. It is quite another to say that reality is contradictory. If it is paradoxical, we can hope for resolution. If it is contradictory, there can be no resolution and science is reduced not only to linguistic confusion but to unintelligibility.

Bohr retreated into an epistemology of contradiction. Had he stopped at the level of paradox, a lot of linguistic confusion could have been avoided.

It is time for scientists to stop speaking in contradictions. They are as confusing as they are meaningless. In struggling to reconcile problems of quantum physics Roger Penrose discusses the problem of light as particle or wave: "How is it that light can consist of particles and of field oscillations at the same time? These two conceptions seem irrevocably opposed."[18]

We notice that Penrose says that the two concepts, wave and particle, *seem* irrevocably opposed. Later he adds: "What does this tell us about the *reality* of the photon's state of existence. . . ? It seems inescapable that the photon must, in some sense, have actually *travelled* both routes at once!"[19] Again he says: "This puzzling feature of quantum reality—namely that we must take seriously that a particle may, in various (different!) ways 'be in two places at once.' . . ."[20]

Penrose's struggle to avoid actual contradiction in his statements is almost tactile. He proceeds with admirable caution, as strained as it may be. He uses the word *seem* as a qualifier. Then he adds other qualifiers. He says that "the photon must, *in some sense,* have actually travelled" and that the "particle may, in various *(different!)* ways be in two places at once."

The words "in some sense" and the "different" (with an exclamation point) enable him to escape contradiction. If he said the particles were at two different places at the same time and in the same relationship rather than "in some sense" or in "different ways," he would slip over the edge. As it is he leaves us with a paradox of phenomena, not a contradiction.

We use similar language in a far more simple sense every day. For example, I might say, "I looked at myself in the mirror this morning." What am I saying? Do I mean that I was both *in front of* the mirror and *in* the mirror at the same time and in the same relationship? Was that really "me" in the mirror? It sure looks like me with some revised images. (I have a mole on my right cheek that looks like it's on the left cheek of the guy in the mirror.) I "appear" to be in the mirror. It is as if I exist in three dimensions outside of the mirror and am somehow recapitulated in two dimensions in the mirror.

Is my mirror image an illusion (like the magician's hat that, although only half empty, appears totally empty when a mirror is used), or is there really something there in the mirror? There is something in the mirror, just as there is something in the mirror of the magician's hat. The reflection may create an illusion, but the reflection itself is not an illusion. There really is a reflection there. What is illusory is the idea that *I*, rather than my reflection, am in the mirror.

I am not suggesting that the paradox of quantum activity can be explained by reflections. (Perhaps some people do. I am not qualified even to speculate about it.) My point is not about light, reflections, or mirror images. It is about language. I say that I "saw myself" in the mirror. This language is an example of the language of appearances, which is not designed to defraud but to speak merely in phenomenological terms. No one faults me for speaking in such paradoxical terms. No one hears me saying that I am in the mirror and not in the mirror at the same time and in the same way. That would be a contradiction, and the contradiction defines the limits of intelligible speech.

Light and *the* Light

"*During one walk Einstein suddenly stopped, turned to me and asked whether I really believed that the moon exists only when I look at it.*"

•

Abraham Pais

We return now to the complexities of speech. We begin with the simple question, "What is a word?" In English we have twenty-six letters in the alphabet. We combine these letters in a huge variety of ways to form discrete units we call words. We use these words to speak and to write. When we write words we use letters. When we speak we use sounds in various combinations. Many different languages emerge using the same alphabet. There are still other languages that use different alphabets. Sounds are more common to broader groups of people than are letters. There are only so many different sounds the human voice can make. Some people master sounds that other people have difficulty with.

The Dutchman struggles if you ask him to say quickly, "Throw those things there." The "th" sound is not a familiar sound in his language. Likewise Americans struggle to imitate certain vowel sounds that are easy for the Dutchman to speak. We do not infer from this that Dutch babies are born with an inability to voice "th" but with an innate ability to make vowel sounds that befuddle us.

The combination of sounds seems to be an acquired ability. By hearing these sounds often enough the infant learns to imitate them.

Apparently cows speak in different languages and make different sounds (sic). In Holland cows say "boe," while in America they say "moo." Not really. Cows don't say either "boe" or "moo." Rather they make sounds we seek to imitate in our own respective languages.

Words are composed of sounds and letters. These words are written or spoken *representations* of ideas. But wait a minute. How can we have ideas without words? Do we not also *think* with words as well as write and speak with them? The written or spoken word uses visual or aural signs and symbols to express ideas.

Then we ask, "What are ideas?" Is there an identity between ideas of reality and reality itself? This is the old question with which Plato wrestled. Are ideas real ontological entities or mere abstract names used to point to reality?

Though we may presently think with words, it is highly unlikely that our thought process *began* with words. How does language actually begin?

I have enjoyed watching not only my own children but my three grandchildren learn how to speak. My three grandchildren, along with their parents, live in our home. The youngest, Michael, only recently learned how to speak. I was involved as a tutor in his learning experience. We played a little game of show and tell. I would point to objects in the room or in pictures and say the name or word for him. I would point to a chair and say, "Chair." Then Michael would mimic my sound and repeat, "Chair." We did this exercise daily. Michael was in the process of associating sounds with objects in his field of vision.

Kant's Apperception

What was going on inside Michael's young mind before he had words to speak with? Certainly he was conscious before he could speak. *Consciousness precedes language.* Of what was he conscious? Philosophers speak of self-consciousness or self-awareness. Immanuel Kant called this, in somewhat esoteric terms, the "transcendental apperception of the ego." Notice he chose the word *apperception* rather than the customary term *perception.* This apperception is further qualified by the adjectival qualifier *transcendental.*

In Kantian terminology the term *transcendental* functions as a sort of practical term. He asked what the necessary conditions are for knowledge to be possible. Since he assigned the self to his noumenal world (the metaphysical realm), the self cannot be perceived. Perception is limited to the world of phenomena.

Kant did not deny the existence of the self. He was agnostic about the ability of the self to be penetrated by theoretical thought. The transcendental apperception then is a kind of necessary presupposition for knowledge. It is not mediated through thought or perception. It is an *immediate* experience, intuitive or self-evident. We might play with this a bit and call it the self-evident self.

How does the self become aware of the external world? Is the external world a mere projection of the mind, a creation of the individual mind? Some have argued this way. The commonsense view, however, disputes this. Realism argues that there really is an external world. You are not a Fig Newton of my imagination, nor am I merely a subject in one of your dreams.

If there is a real world outside of my mind, then any knowledge of it must be either *immediate* or *mediate.* If knowledge of the external world is contained a priori (or innately) in my mind, then all that is necessary for knowledge of it is simple recall or, as Plato supposed (and argued in the dialogue *Meno*), recollection.

If, on the other hand, knowledge of the external world is mediate, then we come by it empirically or through the senses. If this is the case, then our bodies are the *transition* from the self (mind) to the external world.

My avenue to the external world is my senses. I contact the world by taste, touch, smell, sound, and sight. Through this mode of contact I become aware of *sensations.* The baby is aware of sensations. It feels pain and warmth, hears sounds, and responds to visual stimuli.

One of the problems David Hume struggled with is the matter of *sensations.* How do sensations give rise to ideas? Even more basic is the question of how the mind discriminates among sensations.

Science and Taxonomy

Here we encounter the question of *individuation*. If I am bombarded by a host of sensations (the sense manifold), how am I able to sort out these sensations into discrete entities? Are there innate categories or abilities within the mind capable of accomplishing this task? Why, when I open my eyes, do I not see a chaotic blob of sensations?

We might say that individuation is the original and primary task of science. In the biblical narrative of creation the first task given to Adam and Eve was to name the animals. This was the beginning of taxonomy.

Taxonomy is the science of classification. It can be argued that taxonomy is the beginning and end of science, its chief end or goal. It may be an overstatement to proclaim that the *only* function of science is taxonomy, but it is not far from the truth.

The task of taxonomy is the task of individuation. To individuate we must discern similarities and differences among things. The able physician can distinguish between indigestion and stomach cancer, between heartburn and a heart attack. He knows their *similarities* and their *differences.*

The problem of light as a particle or a wave illustrates the problem of individuation. Light seems to possess or exhibit attributes of particleness and of waveness. Particleness refers to the universal class or genus of which every individual particle is a member or species. Likewise waveness refers to the universal class of which each particular wave is a member or species.

Classes are established primarily on the basis of *similarities* among things. Species are established on the basis of *differences* among otherwise similar things. For the definition of anything we take into account both similarity (universal) and dissimilarity (particular). We distinguish between two types of waves on the basis of their dissimilarities but call them both "waves" because of their similarities.

The problem of the wave-versus-particle debate with reference to light is this: Are the categories of waveness and particleness utterly, or at least radically, dissimilar categories? If they are utterly dissimilar, then to assign an entity to both categories is an act of nonsense. If the categories of wave and particle have no point of contact, no aspect of similarity, then it is poor taxonomy to assign something to both categories.

If, for example, light exhibits properties that have similarities to both *waveness* and *particleness* but cannot be conformed to either, then a new category must be assigned. This is, in effect, what happens when people speak of photons as "wavicles." The wavicle serves as a new class of which light is a member. In this case, however, the term *wavicle* cannot properly be defined as something that is both completely a particle *and* completely a wave.

The Paradox of the Trinity

This type of problem is not limited to physicists. It is a problem with which theologians have struggled for centuries.

Christianity rests on two profoundly important and profoundly difficult paradoxes that remain mysteries: the Trinity and the Incarnation. Classically the Trinity was defined in these terms:

God is one in essence
and three in person.

I wish I had a dollar for every time I've heard or seen this formulation described as a "contradiction." Why is it called a contradiction? We are accustomed to thinking in terms of "One person equals one essence." This equation may be a convenient one, but it's not a rationally necessary one. The Trinity is indeed unusual and mysterious, but it is not inherently or analytically irrational.

If the formula for the Trinity asserted that God is one in essence *and* three in essence or that he is three in person *and* one in person, we would be engaging in the nonsense of contradiction. Something cannot be one in A and three in A at the same time and in the same relationship. That's contradiction.

The classical formula of the Trinity is that God is one in one thing (one in A, *essence*) and three in a different thing (three in B, *persona*). The Church Fathers were careful not to formulate the nature of God in contradictory terms. The distinction among persons of the Godhead may be "essential" to Christianity, but the distinction itself is not an essential distinction about God. That is, though the distinction among persons is a *real* and *necessary* distinction, it is not an *essential* distinction.

Lest we seem to be guilty of equivocation here, let me explain further. When I say that the personal distinction among the Godhead is not an essential distinction, I mean by "essential" that which refers to being or essence, not to that which is "important" or "necessary" for other reasons. The distinction is "essential" in the sense that it is important and necessary for our understanding. It is not "essential" in the sense that it distinguishes *being* or *essence* in God.

The formula is not meant to say that essence and person are the same things. Essence refers to the being of God, while person is used here as *subsistence within being.* Essence is primary and persona is secondary. Essence is the similarity, while persona is the dissimilarity in the nature of God. He is unified in one essence, but diversified in three *persona.*

The Paradox of the Incarnation

The same type of problem is encountered in the Incarnation. Here the formula is somewhat reversed. It is said that Christ is *one person* with two natures (essences). He has a divine nature and a human nature. He is one in person (A) and two in nature (B).

Again we have a paradox but no contradiction. The church did not say that Christ is one in A and two in A at the same time and in the same relationship.

Fig. 2
Paradoxes of Christianity

	Essence	Person
The Trinity	God is 1 in essence	God is 3 persons
The Incarnation	Christ has 2 natures	Christ is 1 person

But do we not have a problem here similar to the problem of wave versus particle? I have said that *if* waveness and particleness are mutually exclusive categories it would be nonsense to say that light is both a wave and a particle at the same time and in the same relationship. Am I trying to have my cake and eat it too? Am I allowing in theology what I disallow in physics?

By no means. Godness and humanness are mutually exclusive categories. Something or someone cannot be God and man at the same time and in the same relationship. That is why the formula for the Incarnation is not that Christ is totally God and totally man at the same time and in the same way. We are not saying that Christ's physical body is a divine body. We are saying that the single person has two natures. The divine nature is truly divine, the human nature truly human. The two coexist or are united in one person, but the two natures are not mixed, confused, separated, or divided. Each nature retains its own attributes (see the Chalcedonian Creed). The divine nature is not both divine and human. The human nature is not both human and divine. The person is both human and divine, but not in the same relationship.

We add to the mystery by insisting that the divine nature is not limited to the person of Christ. Deity is not restricted in being to the confines of the human person of Jesus. The finite (human nature) cannot contain the infinite (divine nature). Christ's divine nature is infinite, but his human nature is finite.

It is ironic that, when the church sought metaphors to define the mysterious character of Christ, she chose *light,* calling Christ "the Light of lights." Now we face the same sort of mystery with the nature of light itself. To call Christ the God-man is not altogether unlike calling light a wave-particle. I only hope that physicists will be as careful in formulating the mystery of light as the church has been in formulating the mystery of Christ.

In the taxonomy of theology we seek to pay attention to the problem of individuation via similarities and dissimilarities. In the case of Christ we meet a person who exhibits similarities with God (deity) and similarities with man (humanity). His human nature is dissimilar to his divine nature; his divine nature is dissimilar to his human nature.

We notice that the church did not take the route of the "wavicle." The concept of wavicle suggests a new category that is neither wave nor particle, something that is dissimilar to both though possessing properties similar to both.

Theology rejects the idea of calling Christ a single theanthropic being, a "huvinity" or a "deuman." The reason is that such a theanthropic person would, in the final analysis, be neither human nor divine. Such a view of Christ would be fatal to Christian theology.

To make light a singular being that is a wavicle does not suffer from the same constraints. If such a thing exists as a wavicle, which partakes of properties found in particles and waves but which ultimately is neither wave nor particle, there is no harm to physics.

Combining dissimilar categories to define words is not new in taxonomy. Indeed such a word is encountered every day in science and in the academic world. The word is *universe.* The term *universe* is a mongrel (or bastard) word. It takes two dissimilar concepts, *unity* and *diversity,* and crams them together. The same may be applied to the derivative term *university* as a place where the universe is investigated.

There is an enormous assumption tacitly contained in the word *universe.* The assumption is that reality is both unified and diverse. It assumes an ultimate coherency to the diversity of reality. It assumes cosmos rather than chaos. It assumes the intelligibility of nature, making science possible.

Suppose all we had were diversity. Suppose ours is a multiverse instead of a universe. If all that existed were diverse, could we know anything? How could we distinguish any sensations? If all sensations were diverse, we could never classify them. Taxonomy would be impossible. Individuation would be likewise impossible. We would not even be able to distinguish one sensation from another.

Since individuation depends on similarity as well as dissimilarity, all diversity would become unity. If everything is different, then everything would be the same.

On the other hand if all we had were unity and similarity, a vocabulary of one word would be sufficient. The task of science would be finished in a hurry. The universe would become "unity." This is the problem encountered by crass forms of pantheism. If God is all and all is God, then God is everything. If God is everything, then God is nothing. That is, the term *God* is meaningless because it cannot be individuated from anything else.

Individuation rests upon the ability to distinguish between similarities and dissimilarities. This is how we learn to think and to speak. My grandson's process of learning to speak was a process of individuation. If the beginning of self-consciousness is an awareness of a distinction between the perceiving self and the external world, then the process of individuation begins early on.

To whatever degree the process of individuating objects or ideas of the external world rests on sense perception, to that degree our empirical knowledge is phenomenological.

The Subject-Object Problem

If all our knowledge of the external world rests on perceptions of phenomena, how do we know that the world is as we perceive it? Here we encounter the classic and most basic problem of knowledge and therefore of "science," which simply means "knowledge." This problem is called the subject-object problem.

The subject-object problem has several dimensions to it. The first has to do with the limitations inherent on my sensory powers and abilities. Presently I wear corrective lenses for my impaired vision and am on the edge of needing a hearing aid for my left ear. I have experienced a decline in my visual and auditory senses. Even if my vision were 20/20 and my ears were in perfect working order, I would still have limits. We know that certain animals have more powerful sensory organs than man. Turkeys can see better than we can. Dogs can hear frequency levels we cannot. Deer have superior olfactory powers.

We have been able to expand the horizons of our sensory abilities by developing instruments that aid us. The microscope and the telescope are two such instruments. These perception-enhancers enable us to see things not visible to the naked eye.

When we say that "things are not as they appear to be," in the first instance the statement is elliptical. What is tacit in the statement is the phrase "to the naked eye." The naked eye does not perceive molecules. With perception-enhancers we get a *different* picture of the external world from what we perceive with the naked eye. The new and different picture, however, is still a *perception*. It is an enhanced picture and probably a more accurate picture, but it is still a perception of *phenomena*.

With an enhanced perception of phenomena we are still left with the question of the relationship of the phenomena to objective reality. Is it possible that further enhancements derived by more sophisticated "perception-enhancers" will give us a still different picture of the external world? The Hubble telescope has extended previous boundaries for "as far as the eye can see."

The second dimension of the subject-object problem is the extent to which the perception is in any way influenced by the perceiver. I am not speaking here about the problem of subjective bias by which we "see what we want to see." That is a separate problem. I am speaking about the role that the mind plays in the actual act of perception.

What does the mind do or add to the actual experience of perception? This is the question Hume, Kant, and others struggled with. If it does anything, does what it does affect our perception? If it *affects* our perception, is the effect a distortion of reality? Does the subjective work of the mind enhance our perception of reality or distort it?

This question became a *cause célèbre* in the sixties with the controversy surrounding Harvard professor Timothy Leary's experiments with LSD. Leary was charged with the illicit use of hallucinogenic drugs in psychology experiments. A hallucinogen is so-called because it produces hallucinations. Hallucinations are regarded as distortions of reality.

Leary came up with a provocative defense. He claimed that LSD is not hallucinogenic but "psychedelic." A new word entered into pop vocabulary. A psychedelic drug, argued Leary, is not mind-distorting but "mind-expanding." He paraded witnesses who testified that their perceptions of reality were augmented rather than diminished by the use of LSD. Jazz musicians argued that LSD enabled them to hear tonal patterns they could not perceive without it. Artists declared that their ability to perceive shades of color was intensified by the drug. People even testified that LSD so enhanced their tactile sense that they could experience orgasms in their elbows.

Perception and Reality

The subject-object problem then results in this question: Can I as a thinking and perceiving subject ever get in touch with a real objective world? Stated another way, Do our perceptions correspond to reality?

Reality correspondence becomes a question of truth. The definition of truth itself rests upon the answer to this question. Francis Schaeffer was fond of using the expression "true-truth." He used this expression not because he stuttered or because he was given to tautology. What Schaeffer meant by "true-truth" was "objective truth."

Schaeffer was contending for some correspondence theory of truth. John Locke proposed a correspondence theory by which truth is defined as "that which corresponds to reality as perceived by the senses." This left a bit too much to subjectivity to satisfy George Berkeley. Bishop Berkeley's famous dictum *Esse est percipi* (To be is to be perceived) is often understood to reduce truth to pure subjectivism. If being is determined by perception, then reality depends on the subjective perceiver. This gives rise to such questions as this familiar one: If a tree falls in the woods and no one is there to hear it fall, does it make any sound?

Berkeley sought to escape pure subjectivism by appealing to God as the Great Perceiver. Now the correspondence theory of truth gets a new twist. Truth is then described as "that which corresponds to reality as it is perceived by God." Here room is made, not only for the existence of God, but for the ultimate authority of the revelation of God. If information about reality can be communicated by a perfect perceiver of reality, then we are rescued from both subjectivism and skepticism.

The point of this volume is not to argue for the existence of God or for divine revelation (I have done that elsewhere). Here we are merely canvassing the difficulties that reside in the subject-object problem.

One philosopher, Gordon H. Clark, argues for the necessity of beginning the pursuit of truth with two axiomatic presuppositions: God exists and the Bible is infallible.

Francis
Schaeffer
(1912–1984)

Clark was convinced that any form of empirical epistemology leads inexorably to skepticism. Because of
the subject-object problem Clark insisted that via sense
perception we can never get beyond probabilities. Certainty comes only through reason and the Bible.

Clark's critics jumped on the latter assertion. His
chief critic, John Warwick Montgomery, himself an
ardent defender of Scripture, argued that the only
access we have to the Bible is empirical. We know the
content of Scripture only by reading it (a visual perception), hearing it read (an aural perception), or reading it in braille (a tactile perception).

We recall Albert Einstein's alarm about aspects of
quantum physics. Abraham Pais related a conversa-
tion he had with Einstein in the early 1950s: "We often

Gordon H. Clark
(1902–1985)

discussed his notions on objective reality. I recall that during one walk Einstein suddenly stopped, turned to me and asked whether I really believed that the moon exists only when I look at it."[1]

Stanley L. Jaki sees Einstein's reaction as the development of his earlier concerns about the implications of quantum physics:

It would have, of course, been superhuman to foresee in 1928 such cosmic consequences for a misguided interpretation of alpha-tunnelling as accounted for by quantum mechanics. Even two decades or so later Einstein (and a very few other prominent physicists) still had suspected no such consequences as they deplored the "dangerous game" which the "Copenhagen people" were playing with reality.[2]

The "dangerous game" was a game about reality. It was a contest between science and philosophy, between phenomenology and ontology.

Jaki continues:

> Nor did those in Copenhagen, such as Bohr and Heisenberg, have cosmology in mind as they rejected in the name of quantum mechanics any concern about reality as such, that is, about ontology, as unscientific and unphilosophical, to be avoided at all cost in physics.[3]

Jaki reaches his conclusion about this dangerous game:

> Yet if modern scientific cosmology has reached the point where it may be deprived of its very object, the universe, it is because the champions of the Copenhagen philosophy of quantum mechanics, hardly ever distinguished from the science of quantum mechanics, have succeeded in selling the idea that doing very good science justifies doing philosophy very badly.[4]

Framing the
Question

*"To understand the limitations of science
. . . can be a source of strength,
emboldening us to renew our search for
the objectively real. . . ."*

•

Timothy Ferris

he subject-object problem is a problem about the real. Therefore it is a problem of ontology. If we reduce the problem by allowing one pole to be swallowed by the other, we destroy science. If the objective is swallowed by the subjective, science becomes merely a matter of psychology. If the subjective is swallowed by the objective, we are left with a chaos of sensations.

Timothy Ferris comments on this nagging problem in *The Mind's Sky*:

> I envision our relationship to the universe as symmetrical, hourglass-shaped. On one side is the outer realm, inhabited by galaxies, stars, the plants and animals, and our fellow human beings. Most of us (the solipsists aside) believe that this outer world exists, though we appreciate that our direct perceptions of it are limited and skewed. On the other side is the inner realm of the mind, where each of us is destined to live and die; here resides all we can ever know. Through the neck of the glass flow the sense data by which we perceive the outer realm, and (flowing the opposite way) the models and concepts we apply to nature, and the alterations and abridgments we impose on her. [See fig. 3.] We tip this imaginary hourglass from time

to time. In the nineteenth century, when classical physics ruled, we tended to think of the sand as flowing almost entirely from the outer to the inner realm, from an objectively real world to our passively recording minds. In the twentieth century the concept of observer-dependent phenomena in quantum physics has shifted our attention to the ways our observations influence how we perceive nature. But so long as there are thinking beings in the universe, neither bulb of the hourglass will ever be empty.[1]

Ferris's Hourglass and Magritte's Pipe

Ferris's hourglass speaks of "observer-dependent phenomena." Is this a modern retreat to George Berkeley's *Esse est percipi*? Do the phenomena not exist apart from the observer? This suggests a subject-object problem with vengeance. It is one thing to say that the accuracy of our perceptions depends on the acuteness of our observations. It is quite another to say that the perceptions themselves are dependent on the observer. Ferris doesn't explain in what sense the perceptions are dependent on the observer. If they are totally dependent, how can we escape the solipsism he apparently eschews by setting "the solipsists aside"?

The hourglass analogy itself suggests that Ferris means, not that perceptions depend totally on the observer, but that observation is a "frame" in which perceptions are received. Ferris later introduces a critical-frame theory to his treatment of the subject.

Fig. 3

Ferris's Hourglass

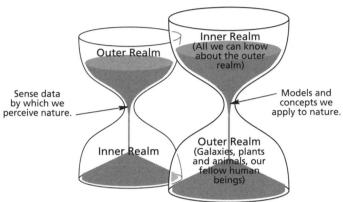

He recounts a fascinating episode from the world of art.

Ferris tells about a painting of a smoker's pipe painted by René Magritte in the 1920s: "A pipe, depicted with photographic realism, floats above a line of careful, schoolboy script that reads, *Ceci n'est pas une pipe*—'This is not a pipe.'"[2]

What did Magritte intend by the inscription? Was he merely pointing to the obvious, that a painting is not the reality it portrays but merely a representation of it? Ferris doesn't think so. He alludes to later modulations of Magritte:

In *The Air and the Song*, painted in 1964, just three years before Magritte's death, the pipe is found inside a representation of an elaborate, carved frame, as if to emphasize that it is only a painting—yet smoke from its bowl billows up out of the painted "frame"! In another canvas, *The Two Mysteries*, Magritte is even more insistent: The original pipe painting, complete with caption, is depicted as sitting on an easel that rests on a plank floor; but above it to the left hovers a *second* pipe, larger (or closer) than the painted canvas and its frame. What we have here is a painting of a paradox. Obviously the smaller pipe is a painting and not a pipe. But what is the second pipe, the one that looms outside the represented canvas? And if that, too, is but a painting, then where does the painting end?[3]

Magritte's mystery is not about pipes; it is about frames. Ferris is talking about the function of frames or frameworks in our perceptions of reality.

Some assume that everything outside the frame is not real. The pipe is not a pipe until it appears within a frame. Ferris adds:

The enemy of surrealists like Magritte, and of artists generally, is naive realism—the dogged assumption that the human sensory apparatus accurately records the one and only real world, of which the human brain can make but one accurate model. . . .

The truth, of course, is that nobody can grasp reality whole, that each person's universe is to some extent unique, and that this circumstance makes it impossible for us to prove that there is but one true reality. . . . *Everything* thus is framed, cut from its cosmic context by the limitations and peculiarities of our sensory apparatus, the prejudices of our presuppositions, the multiplicity of each individual mind, and the restrictions of our language.[4]

Photograph by Paul Bijtebier

The Air and the Song (1928)
by René Magritte

Photograph by Archives Galerie Iolas

The Air and the Song (1964)
by René Magritte

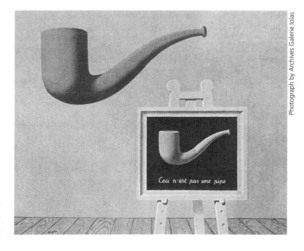

Photograph by Archives Galerie Iolas

The Two Mysteries (1966)
by René Magritte

• • 105 • •

At first blush it seems as if Ferris is siding with the solipsists. Statements like "Each person's universe is to some extent unique" suggest that there is no single, objective reality. It seems to suggest that there are as many universes as there are persons or perceivers. Thus, when I invite a friend over to dinner, I am inviting him to participate in *my* universe.

At second glance, however, this does not seem to be what Ferris means. Perhaps he means that every person has his own unique frame or perspective of the universe. He says "*Everything* . . . is framed." He notes that this is related to the restrictions of our language.

The framing subject, as he deals with phenomena, imposes his frame not only on the phenomena, but on the language he uses to describe or represent the phenomena. Herein lies one of the most severe difficulties inherent in the subject-object problem.

When I speak, I speak in phenomenological language. That language is conditioned by my personal, subjective frame. If we learn language a posteriori, out of our own experience, and if no two people have exactly identical experiences and exactly identical frames, then this will affect the language by which we communicate.

If, for example, my experience of dogs includes a nasty, traumatic encounter with a vicious pit bull, this may color my perception of what a dog is. The word *dog* now functions in my memory bank with a highly charged, pejorative nuance.

Because no two persons' experience is identical, words we use to communicate are not totally and absolutely *univocal.* We can approach and approximate the univocal, but because of the diversity of our "frames," our language is always tainted to some degree by the equivocal.

God as "Wholly Other"

We encounter this problem in theology. I once had a discussion with a group of theologians who insisted that God is "wholly other" (*totaliter Aliter*) from man. I asked them how they knew anything about God. They quickly responded, "Because he reveals himself."

I replied, "Perhaps I haven't made my question clear. *How* does this God reveal himself?"

They responded, "Through nature, the Bible, and through history."

Again I persisted: "My question is more fundamental. If God is *totally* or *wholly* dissimilar to us, how can he communicate anything to us about himself through any means? Unless there is some point of similarity, I cannot see how any intelligible communication is possible."

They thought for a second and said, "Maybe we should not say that God is wholly other."

The same problem exists in intelligible communication about anything between any two persons. If everybody's frame or everybody's universe were *utterly* unique (though uniqueness does not admit to degrees), there could be no intelligible communication.

When Ferris says that "each person's universe is *to some extent unique,*" I assume he means that each person's universe or "frame" is to some extent *dissimilar* to everyone else's. If it were totally dissimilar, the same problem that applies to "God-talk" applies also to "man-talk."

We are back to the question of taxonomy. Taxonomy requires both *similarity* and *dissimilarity* to work. This underscores the importance of *analogy* to human communication. It is similarity that makes analogical language possible and meaningful. It is dissimilarity that makes *analogy* fall short of *univocality.*

That Ferris does not surrender to despair or subjectivism may be seen in this statement: "To understand the limitations of science (and art, and philosophy) can be a source of strength, emboldening us to renew our search for the *objectively real* even though we understand that the search will never end."[5]

Magritte's painting of the pipe in the first instance is a *representation* of a real pipe. Subsequent paintings of the painting within a painting pose the paradox of representation. Each step of representation is one step further removed from reality.

As a boy I was mesmerized by mirrors at the barbershop. As I sat in the barber chair, I could see my image in the mirror in front of me. The shop also had mirrors on the wall behind me. In the mirror ahead I could see my image plus the reflection of that image on the mirrors behind. The mirror behind reflected the image from the mirror ahead. Each discrete image was smaller and smaller, until the multiple reflections vanished at the horizon point. I used to sit in the chair and try to count the number of images I could see in the back-to-back reflections. The experience always ended in frustration, with my losing count along the way.

As the painting represents the pipe, so my images in the mirror represented me. The problem of progressive removal from the thing itself plagues our attempts to get to the reality.

If we define phenomena as the *appearance* of reality, we can easily substitute the word *representation* for appearance, translating into *phenomena* as the representation of reality.

This does not necessitate the inference that the phenomena themselves are not reality. Aristotle's distinction between substance and accidens applies.

With later empirical philosophy the distinction between essence (substance) and phenomena (accidens) was reworked. John Locke distinguished between substance (or essence) and qualities. He further distinguished between two *kinds* of qualities, *primary* and *secondary* qualities. Primary qualities, like Aristotle's *accidens,* were inherent in or at least necessarily and ontologically *linked* to substance. Secondary qualities were added by the mind in conjunction with other elements.

For example, we might describe color as secondary rather than primary. We might ask a simple question, "What color is a lemon?" The normal answer, of course, is "Yellow." Then we pose the question, "What color is a lemon when the lights are off?" Without light lemons have no color. They appear yellow in the light because all the light waves of the spectrum are absorbed by the lemon except the yellow band, which is reflected. The lemon "receives" its color from light. Yellowness is not intrinsic or inherent to lemons. It is a secondary quality.

On the other hand, primary qualities or accidens seem to be inherent in substance in the sense that there is a necessary connection between a thing and its peculiar qualities. (This was disputed by both Bishop Berkeley and David Hume.)

The Doctrine of Transubstantiation

The problem of the relationship between the essence of a thing (*Ding-en-sich*) and its accidens or qualities has been the source of endless controversy in the church. The Roman Catholic doctrine of transubstantiation has been the focal point of the debate. Though cloaked in Aristotelian terminology, the church breaks sharply at one crucial point with the philosopher.

The formula for transubstantiation is this: In the miracle of the Mass the *substance* of bread and wine *changes* (trans-, moves "across") into the substance of the body and blood of Christ, but the *accidens* of bread and of wine remains the same.

As a result of this miraculous change we have the substance of one thing present without its corresponding accidens, and the accidens of another thing without its corresponding substance. We have the substance of Christ's body and blood and the accidens of bread and wine.

Martin Luther protested against this because it involves *two miracles,* which he thought was a frivolous and unnecessary philosophizing of the Mass. Two miracles are required: one to have the substance of a thing present without its accidens, and another to have the accidens present without its substance.

The idea of a miracle is introduced here to "save the phenomena" and to account for the departure from Aristotle's view that substance and accidens are immutably joined together. What Aristotle joined together, the church rent asunder.

The phenomena had to be "saved" because, before the miracle takes place in the Mass, the common elements of bread and wine still have the accidens of bread and wine. They are clearly *perceived* as bread and wine. They look like bread and wine, taste like bread and wine, smell like bread and wine, and feel like bread and wine. If we dropped them, they would sound like bread and wine.

After the miracle of transubstantiation there is a change in reality but not in perception. It is an *unperceived* transformation. The essence may change but the *phenomena* remain the same. The elements still taste like bread and wine, look like bread and wine, and so forth.

There are other theological issues that make transubstantiation problematic. Our concern at this point, however, is not primarily theological but epistemological and ontological. That Rome sees a miracle as special supernatural work necessary for transubstantiation indicates a tacit commitment to a view of nature or science that sees a natural and real link between substance and accidens.

We may define *accidens* or *primary qualities* as the external manifestation of substance. It is more than representation, as in the case of a painting or a word. The primary qualities are themselves an integral part of reality.

If that is so, then when we are in touch with the phenomena, we are in touch, at least "accidentally," with reality. There may be a deeper or hidden dimension to the reality that remains unperceived, but this does not warrant the inference that there is no reality underlying the phenomena.

In traditional physics the link between essence and phenomena was assumed to be a causal one. This is why the question of causality burns so intensely in the contemporary debate. This issue is so germane to the reconciliation of quantum physics with classical physics that we will devote more attention to it later.

Our problem with the representational aspect of phenomena to essence is exacerbated when the categories of secondary and tertiary qualities are added.

If some qualities are "added" to reality by the mind of the observer, how can we ever overcome the subject-object problem? How can we assign *reality* to anything?

The Reality of Our Perceptions

This raises the question of the "reality" of our perceptions. If the phenomena contain both objective qualities (primary) and subjective qualities (secondary or tertiary), how can we ever discern the difference between reality and illusion?

One thing we can do is, by cutting the Gordian knot, join the emboldened focus of Ferris and "renew the search for the objectively real even though we understand that the search will never end." This was the passion of Allan Bloom, whose book *The Closing of the American Mind* was, most unexpectedly, a runaway best-seller. Bloom complained that the "what" to which the modern mind was being closed was *objective truth.* To retreat into philosophical relativism and skepticism was not an option for Bloom. He rightly foresaw that such a retreat would be the death knell to scientific progress and achievement.[6]

If I may indulge in moralizing, let me say that any fool can be a skeptic. Cynicism and skepticism are the crudest forms of quasi-intellectualism. The skeptic assumes an Olympian posture above the realm of lesser mortals. He remains "above" the agonizing pursuit of truth, content to be regarded as superior to the fray. Let the cynic become cynical about his cynicism and the skeptic skeptical of his skepticism and join the battle.

It takes courage to sail our ships into uncharted waters, as Friedrich Nietzsche declared. It is the courage and determination of the explorer, of the Magellans of this world, to press ahead in exploring the unknown.

We do not require an existential leap of faith to pursue the quest. There is reason for the engagement. So many frontiers have been conquered, so much has been learned, that it is not a fool's errand to continue the quest. As my former bridge partner (who annoyed me by constantly overbidding his hand) used to say, "A faint heart never won a fair lady."

Beyond such Gordian-knot cutting there is something else we can do. We can make crucial distinctions between different *levels* of reality. Primary qualities or accidens may not be the deepest or ultimate level of reality, but that does not mean they do not exist or are "unreal."

In philosophy, at the level of metaphysics, a distinction is made between essence and existence. This distinction refers to modes of reality. It takes its cue from the ancient Greek distinction between *being* and *becoming.* It has been said that the history of philosophy is nothing more than a postscript to the writings of Plato and Aristotle. Surely this is hyperbole. We can extend the hyperbole a step further by saying that the history of philosophy is nothing more than a postscript to Parmenides and Heraclitus.

Little survives of Parmenides' works save allusions and quotes in other ancient writers. The most famous quote attributed to Parmenides is the apparent redundancy, "Whatever is, is." The first time I heard that I mused, "And this guy is famous?"

Yet no statement I've ever read in philosophy has haunted and nagged me more than this citation from Parmenides. He was making a statement about being or essence.

Parmenides was challenged by Heraclitus, who said that everything is in a state of flux. His counterdictum was, "Whatever is, is changing." He added the assertion, "You cannot step into the same river twice." The river is moving. It is changing. We are changing as we move from step to step. If nothing else changes, the constant variable we deal with is time. Each experience I have takes place in a different time frame.

To restate Heraclitus's views we would say that whatever is, is becoming. Herein lies the classic distinction between being and becoming, essence and phenomena, actuality and potentiality.

The problem of being and becoming may be stated this way: "If all is being, then being is actually everything and potentially nothing." Pure being has no becoming, no change, no potential.

On the other hand pure becoming would be potentially anything but actually nothing. This conundrum leads some philosophers to assert that if everything is changing, then nothing exists.

Existence then becomes [sic] the operative word. What do we mean when we say things really exist? Existence can refer to a mediate position between being and nothingness (nonbeing). The word *existence* derives from the Latin *existere,* which means literally "to stand out of." Existence is something "outstanding." It "stands out" of something. Existence is linked to being and essence as that which stands out of being and nonbeing. It sort of hangs suspended between being (essence) and nonbeing (nothing). To exist is to participate in being. To exist is to have one foot in being and one foot in nonbeing (a linguistically problematical phrase, to be sure).

We see the linguistic problem when it comes to defining the difference between God and humans. In popular jargon people are called human beings, not human becomings. God is referred to as the Supreme Being. What is the difference? The difference is in *being.* God is seen as pure or self-existent being, the one who has the power of Being in himself. There is no "becoming" in God. In Aristotelian categories God is Pure Actuality.

Human beings change. They are in a state of becoming, manifesting elements of potentiality. We say, in philosophical language, that human beings exist, but at a lower level or mode of being than God. They "stand out" of being but are not nothing.

In this sense God does not "exist." That is, he does not stand out of or under being; he is Being. In metaphysical terms God is, but he is essence, not existence. We exist, but we are not pure essence.

On the other hand, in ordinary language when we say that something exists, we mean that it is real. It is something, not nothing.

These metaphysical distinctions become important when we turn our attention to *phenomena.* The phenomenal may not be the level of essence, but this does not mean that phenomena have no existence. Phenomena are real if reality extends beyond the level of essence to the level of existence.

Though Immanuel Kant was skeptical about our ability to reach the noumenal realm of essence, he was not skeptical about reaching the realm of the phenomenal. For science to explore the realm of phenomena is not to limit exploration to nothing. This is not a case of much ado about nothing. The phenomenal realm is the realm of existence and therefore at least one level of reality.

We can pursue the idea of levels of reality by distinguishing between *ultimate* and *proximate* reality. Proximate reality is the level of existence. Ultimate reality is the level of essence. Ultimate reality may also be called the *remote* level of reality, while approximate reality refers to the *near* level of reality.

Illusions and Relativism

What about illusions? Illusions refer to phenomena that are linked differently from essence. With illusions we have perceptions of things that have no underlying essence or reality. The desert traveler suffers an illusion when he sees the mirage of an oasis. The oasis does not exist. But something exists, namely the mirage or the illusion. If there is such a thing as causality, then something must exist to produce the illusion. Here the illusion finds its home or cause, not in objective, external reality, but in subjective, internal perception.

The illusion has to come from somewhere or something. Though the illusion is not a picture of reality, it betrays the phenomena of reality.

Let us illustrate this by a common phenomenon in human experience. Two people respond to the same event with disparate perceptions and with different feelings or emotional reactions. Recently my wife was involved in a minor traffic accident, a fender-bender. Her car was hit in the side by a truck. The truck driver insisted that it was my wife's fault, and my wife insisted it was the trucker's fault. A third-party eyewitness testified to the attending police officer that my wife's version was correct. After interrogating the parties and examining the physical evidence, the policeman cited the trucker for negligence.

When the trucker gave his version of the accident, he was not necessarily lying intentionally. He saw the event through his frame. It is possible that he really believed that things happened as he perceived them. In this case we have different levels of reality here. The event can be broken down into the following constituent aspects:

1. What actually happened
2. The parties' perceptions of what happened
3. The parties' feelings about what happened

Now one or more of the parties' perceptions of what happened may not correspond to what actually happened. Yet the perception may be a *real* perception. Its incorrectness does not vitiate the reality of the perception. The perception was not a perception of reality. It was a real perception of unreality.

Such is the nature of illusion. An illusion is a mistaken though real perception of reality. This happens every day apart from hallucinations, delirium, mirages, and the like. When I am called to mediate arguments between people, I usually point out that though the people disagree about what the truth is, this does not negate the truth that both parties really believe or perceive that they are right.

We do not solve the dispute by a retreat into the nonsense of relativism. If the argument is about slamming a door, the options are two. Either the door was slammed or it wasn't. Both disputants cannot be right. If I slammed the door and perceive that I didn't slam the door, all my subjective belief within my subjective frame does not alter the objective reality. We don't resolve the dispute with a denial of objective reality . . . subjectivizing or relativizing it.

We can ameliorate the feelings about the dispute by granting that there is another reality involved, namely the subjective perception. How can that help?

If my wife sees me slam the door and becomes angry at my insensitivity in slamming the door, her anger is raised a notch if I add insult to injury by insisting that I did not slam the door. Now, in my wife's eyes I am not merely an insensitive door-slammer, but an insensitive and *lying* door-slammer. If my wife can be persuaded that I really believe I did not slam the door, she is mollified at least to the degree of her original anger, retreating to her conviction that I am an insensitive door-slammer but an honest one.

If the argument proceeds and I convince my wife that *if* I slammed the door I did not do it intentionally, then perhaps she can take refuge in my being a sensitive door-slammer. If I become convinced that I actually *did* slam the door and apologize for it, we can make a great leap forward to reconciliation.

What doesn't help is to deny that the door exists or, even worse, to insist that the door slammed by chance. Either of these involves an excursion into the absurd.

If phenomena are representations of essential reality and if phenomena are themselves real, then phenomena may be considered as real *avenues* to essence.

Imagination and Virtual Reality

If phenomena do not reside strictly in my subjective mind, then science is possible. The external world is not merely the product of my imagination.

We will explore the imagination more fully later. For now let me illustrate the role of imagination in the perception of phenomena by another reference to my grandson Michael. Recently Michael's father was reading to him at bedtime from a children's Bible-story book. After finishing the story about Jesus' walking on the Sea of Galilee, Michael's father asked, "Michael, how do you suppose Jesus was able to walk on the water?" Michael thought for a moment and replied, "He used his imagination!"

The power of imagination gives us more than a platform for make-believe. The imagination can be useful in the creative enterprise as well as in problem-solving brainstorming. With the imagination we can *suppose* reality that may in fact correspond to actual states of affairs.

The fascinating development of virtual reality gives us a glimpse of the broadening horizons of phenomenological perception.

If phenomena belong to a secondary level of reality, how can they be useful to us in our quest for knowledge? Ferris writes: "Virtual reality (VR), the latest development in computer-sensory interfacing, significantly deepens one's immersion in the computer-generated simulation."[7]

The use of computer technology now makes it possible to replicate real-life situations complete with three-dimensional imagery. It will be possible to "enter" future films and be part of the surroundings. It will be "the next best thing to being there."

Ferris himself experienced a VR simulation by which he spent one-and-a-half hours "exploring" the surface of Mars:

> I put on the helmet and found myself standing on a rocky promontory, looking down across a jumble of cliffs and plateaus stained in unearthly shades of ocher, sand-yellow, and plum. Using the computer controls I could descend to the valley floor and wander for miles through the twists and turns of one of the solar system's most imposing landscapes, or climb high into the

pink sky and take in the wider view, studying inky bluffs that marched off toward distant peaks five hundred kilometers away. . . .

And despite its limitations, the experience of walking on Mars was vivid and immediate, not at all like seeing a movie or a photograph. The way I remember it, I was *there.*[8]

In Ferris's vivid description of his VR experience he sounds like a living Super Mario. If he ever does have the opportunity to visit Mars, it will be an occasion for déjà vu.

VR technology provides the most advanced representation of reality imaginable. Ferris stresses the contrast between VR and a movie or photograph. His experience, he says, was *immediate.* From the perspective of *memory,* there is no difference between having been on Mars and having experienced it via VR.

However, the technology is not called simply R. It is *V*R, or *virtual* reality. There is still a clear distinction between reality and "virtual" reality. The qualifier *virtual* indicates that the experience was *almost,* but not quite, the real McCoy.

Though Ferris described his experience as "immediate," in fact it was a *mediated* experience. Though the technology so vividly replicated the reality of Mars, it was still a reproduction or phenomenological *representation* of Mars. The vivacity of the representation was so intense that the observing subject (Ferris) could not distinguish between the real object (Mars) and the virtual representation of Mars.

Words Revisited

Earlier we considered the representational dimension of language. Words represent objects (nouns), actions (verbs), and other relationships (conjunctions, prepositions, etc.). The word *tree* represents the actual tree or my phenomenological perception of tree. The symbol or sign-representation of the word *tree* becomes a concept in my mind.

There is a cyclical reciprocity between awareness and words. I begin with an awareness of a tree. I experience the phenomenon of a tree (or a picture of a tree). I then learn to associate the word (sound or written symbol) *tree* with an object in the external world. When I am in a room by myself, I can now think of the word *tree* and have a mental concept of what the word represents. The word can now trigger an *awareness* of trees. The cycle goes something like this (see fig. 4):

Fig. 4

Awareness and Words

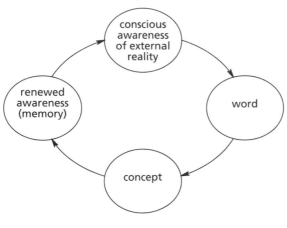

conscious awareness of external reality leads to *word,* which leads to *concept,* which leads to *renewed awareness* (memory).

We encounter this cycle in the study of theology. We begin with a vague sense or "awareness" of God. Theologians describe it as the *sensus divinitatis,* and they ascribe it to *immediate* general revelation. It is argued that such an awareness of the deity is innate or a priori to the soul of a human being. Friedrich Schleiermacher called it "God-consciousness."

From the initial awareness of God comes the attempt to conceptualize him via words. The study of doctrine is a study *about* God. The Bible is not God himself. To learn the Bible is to learn about God. The hope in theology is that a knowledge of words or ideas about God will reawaken the soul to an awareness of God himself.

Another way of saying this is that theological words point beyond themselves to a Being. The words represent the Being, but are not the Being himself. Insofar as words are representations that point beyond themselves to another reality, they belong to the category of phenomena.

Words are an avenue to reality. To be sure, words can be empty, meaningless, or unintelligible. They can also be content-laden, meaningful, and intelligible.

When words are grouped together into propositions, their intelligibility can be measured formally. This is the task of reason.

The Policeman of Science

"My first mistake was in having assumed that the orbit on which planets move is a circle. This mistake . . . had been supported by the authority of all the philosophers. . . ."

•

Johannes Kepler

As thinking subjects we are bombarded by a myriad of sensory stimuli. The mind orders perceptions into categories. The mind does the work of taxonomy. We use thousands upon thousands of words to shape our ideas and concepts. Our ideas are then shaped into propositions.

If it is the function of the mind to seek or to find order in the phenomena, then it is the function of reason to arbitrate that order.

We use the term *reason* in various ways. Sometimes it functions as a virtual synonym for *cause.* We say that A is the reason or the cause for B. Sometimes the word refers to our entire *process* of thinking. Here it is a virtual synonym for *cogitation.* At other times *reason* is used as an antonym of *faith,* to refer to what is "known" as distinguished from what is "believed." Other uses of the word *reason* could also be delineated. For now, however, our chief concern is for the formal function of reason, which is tied to *logic* or *deduction.*

Earlier we pointed to deduction as one leg of the two-legged support system of science, the scientific method itself. Here we distinguish between the gathering of data empirically and phenomenologically as *induction,* and the influences we draw from the data as *deduction*. It is at the point of deduction that logic functions as the governor or policeman of science.

We also noted earlier that logic had been defined by Aristotle as the *organon* of all science, the necessary "tool" or "instrument" for all meaningful discourse. Again we assert that Aristotle did not invent logic; he merely "discovered" it or formulated its basic structure.

Laws of Logic

In the history of theoretical thought other systems of logic have been defined that differ at various points from Aristotelian logic. Yet there are certain *laws* of logic that apply universally. We have already noted the primary and foundational importance of the law of noncontradiction. Insofar as this law is a law of deduction and insofar as deduction is integral to the scientific method, we can say that the law of noncontradiction is a law of science. It is certainly a law of science in the broad sense of *science*. That is, it is a law of *knowledge*, because where the law is violated no knowledge or intelligible discourse is possible.

In addition to the law of noncontradiction, logic also provides us with the *laws of immediate inference*. The laws of inference set the parameters for legitimate deductions made from data. The "logic of the facts" for which Enlightenment thinkers sought cannot be found if the laws of immediate inference are violated.

The laws of immediate inference specify various categories of propositions and define their relationships. For example we have universal affirmatives, universal negatives, particular affirmatives, and particular negatives (see fig. 5).

The laws of immediate inference govern the logical relationships among these categories. For example, the following pairs of propositions, when each pair is taken together, violate the laws of immediate inference:

A. All men are mortal.
B. No men are mortal.
   or
A. All men are mortal.
B. Some men are not mortal.

Fig. 5
Categories of Propositions

	Example*	Explanation
Universal affirmative	All men are mortal.	Something is affirmed (mortality) about every member of a class (men).
Universal negative	No men are Martians.	Something is denied (Martianness) about every member of a class (men).
Particular affirmative	Some men are bald.	Something is affirmed (baldness) about particular members of a class (men).
Particular negative	Some men are not Republicans.	Something is denied (Republicanness) about particular members of a class (men).

*The point is not to debate the truth or falsehood of these statements, but to classify them.

or
A. No men are bald.
B. All men are bald.
   or
A. No men are bald.
B. Some men are bald.

With each of these examples logic blows its whistle and cries, "Stop in the name of the law!" The policeman does not determine which, if any, of the individual propositions are true. He merely issues a citation for violating the rules of thought and intelligible discourse.

When we get to particular affirmatives and particular negatives, the violations are not always as easy to discern. If we say "Some squirrels have bushy tails," it is possible that all squirrels have bushy tails but we haven't yet reached that generalized or universal conclusion. If we examine ten million squirrels and discover that all of them have bushy tails, we may be tempted to make a universal judgment about squirrels. We know, however, that tomorrow we may find a squirrel without a bushy tail. Then we will revise our universal and say that *some* or even *most* squirrels have bushy tails.

But here, via induction, we are concerned about the truth value of the proposition "*All* squirrels have bushy tails." This is an inductive question. It becomes a deductive question when we set the initial proposition alongside another proposition:

A. All squirrels have bushy tails
B. Some squirrels have bushy tails.

In this instance both propositions can be true. If it is true that all members of a class have a particular attribute, then it is a valid inference that *some* of them have this attribute.

Another relationship may be:

A. Some men are *bald.* (particular affirmative)
B. Some men are not *bald.* (particular negative)

Here we have a more confusing set of relationships. The two propositions, when taken together, could both be true. They could also not be true. We have seen that it's possible to conclude tentatively that some A is B when in actuality all A is B.

Though there is a *possible* harmonious relationship between a particular affirmative and a particular negative, the relationship is not *necessary.* If it is true that some men are mortal, it does not necessarily follow that some men are not mortal. The statement "*Some men are mortal*" *implies* that some men are not mortal, but this inference is not demanded as a logical necessity. Here we see the difference between a *possible inference* and a *necessary inference.* The first one logic allows as a possibility, the second one logic demands.

Truth and Validity

In terms of logical analysis we say that propositions may be true or false, meaningful or meaningless. Arguments, however, are not so judged. Arguments are neither true nor false; they are *valid* or *invalid*. Let us examine the following syllogism:

A. All moons are made of green cheese.
B. The object that illumines our nights is our moon.
∴ Our moon is made of green cheese.

Is this a valid argument? Yes. The conclusion follows from the premises by resistless logic. Is the conclusion true? Few of us, if any, believe that it is.

What logic tells us here is that *if* all moons are made of green cheese and *if* the thing that appears in our sky is a moon, then it is absolutely certain that the moon is made of green cheese. The composition of the moon is not a question of logic; the argument *is* a question of logic.

Logic, in and of itself, has no content. It monitors the relationships among propositions. It cannot verify the truth of a proposition (except analytically or formally, as we will see). It can verify the validity of an argument.

Perhaps logic's most valuable contribution is at the level of falsification. Here, like Teddy Roosevelt, logic can "walk softly and carry a big stick." Logic can demonstrate conclusively that an argument is invalid. It can also show a proposition to be formally false. For example, the statement "A bachelor is a married man" is analytically or logically false. A bachelor cannot be unmarried and married at the same time and in the same relationship. He can be "married to his work" and unmarried to a woman at the same time. But he cannot be married and unmarried to his work or to a woman at the same time and in the same relationship.

When statements are analytically false or arguments are inferentially invalid, logic works like a tilt light on a pinball machine. It is like the talking computer that sounds a buzzer and squawks, "Does not compute!"

The great service logic performs in our quest for truth is that it alerts us to error. Finding the error in our thinking or computation is vital to our quest for a correct analysis of the data.

Challenging Assumptions

I once heard a man define his "law of creativity." His law was simple: "Challenge the assumptions." He explained it this way: "If you want to have a creative Christmas tree, challenge the assumptions of Christmas trees."

One assumption is that Christmas trees are green. Paint it red. Another is that they are pine trees. Select a palm tree. Another is that they have glass ornaments. Adorn it with bananas. (I have a friend who had a hook fastened in his family room ceiling when his house was built. Each year he hangs a Christmas tree from the hook, upside down. *That* is creativity.)

This principle is frequently applied in the arts and provokes changes in the direction of music, painting, and the like. But what about science? In decorating a Christmas tree we are not seeking the "truth" or objective reality about Christmas trees. Natural science is not seeking for creativity. It is seeking for progress, for better paradigms to account for the data.

We may challenge the assumptions about Christmas trees willy nilly and be creative with no harm. In science, however, challenging the assumptions will hardly profit if it is done willy-nilly. To challenge the assumptions in the scientific realm is to search for assumptions that may be erroneous or misleading. It is the type of work done by detectives like Sherlock Holmes.

We use this technique in solving certain mind games or riddles. My family and I like to play a game we call "How's come?" This game involves a certain stated scenario that requires an explanation. For example:

Two men are found dead in a cabin in the woods. How's come?

The task is to guess the full story from the brief scenario offered above. The player is allowed to ask questions that can be answered only by yes or no: Were they shot? Did they die of a disease? And so on.

The answer to this How's come? problem is that the two men were killed in a plane crash. Their bodies were discovered in the woods. They were trapped in the cabin of their plane.

The key to solving such puzzles is to challenge the assumptions that are usually concealed in the clues. When someone speaks of a "cabin in the woods," we are inclined to assume that it is a small dwelling place, not the cabin of an airplane.

My favorite How's come? is this little ditty:

A bell rang.
A man died.
A bell rang.
How's come?

This puzzle has more than one hidden assumption. To determine the cause of the man's death is fraught with difficulty because the information is so scant and misleading.

To solve it we have to consider various kinds of bells. We must not assume that both bells are the same bell or even the same type of bell. We must not assume that the man who dies is a normal man.

Shall I let you solve it? That would be nasty. You would not even have the benefit of yes-or-no answers to your questions.

Okay. I won't be nasty. Here is the scenario. A man, who is blind, goes swimming in the harbor near his home. He keeps his sense of direction toward land by listening for the church bell that rings every quarter-hour. He swims out into the harbor and gets a distance away from the church bell. He hears a bell ring. It is, sadly, a buoy bell far removed from shore. He thinks it is the church bell and so swims toward the sound. It is beyond his reach. Instead of swimming to the safety of the shore, he drowns by swimming too far out to sea. After he drowns the church bell rings. Too late.

Once we challenge the assumptions that the man can see and that both bells are the same, the conundrum is easily solved.

Once a person hears a few How's come? problems, he can quickly become proficient in solving all such tales he hears because he learns the technique of challenging the assumptions.

Not all the assumptions are erroneous. The trick is in finding the erroneous one(s). Once an erroneous assumption creeps into the game, it has the nasty habit of perpetuating itself. False assumptions are hard to kill. They tend to have nine lives.

The Assumption of Circular Orbits

Consider the assumption of the circularity of planetary orbits. This thesis went back to the Pythagorean number theory that attached mystical significance to numbers and to the perfection of the sphere. This assumption persisted through Plato, Aristotle, and Ptolemy.

When Nicolaus Copernicus provoked his revolution, the revolution was still tied to the circle. Timothy Ferris notes:

> Trouble arose not in the incentive for the Copernican cosmology, but in its execution. (The devil, like God, is in the details.) When Copernicus, after considerable

toil, managed to complete a fully realized model of the universe based upon the heliocentric hypothesis—the model set forth, eventually, in *De Revolutionibus* [see

Fig. 6

Copernicus's Model of the Universe

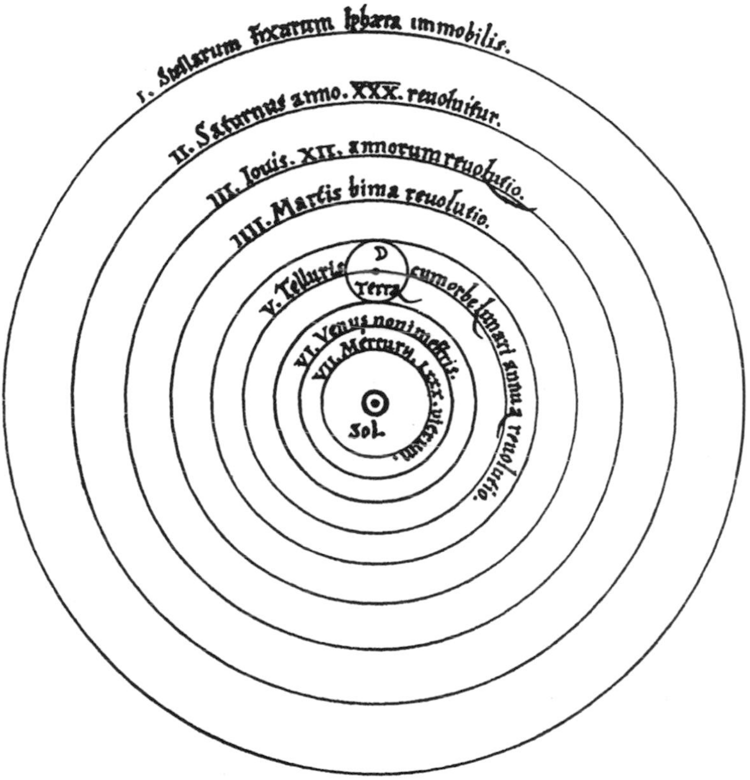

I. Immobile sphere of the fixed stars
II. Saturn completes one revolution every 30 years
III. One revolution of Jupiter every 12 years
IV. Biannual revolution of Mars
V. Annual revolution of Earth and sphere of Moon
 Terra = Earth
VI. Venus every 7.5 months
VII. Mercury in 88 days
 Sol = Sun

fig. 6]—he found that it worked little better than the
Ptolemaic model. One difficulty was that Copernicus,
like Aristotle and Eudoxus before him, was enthralled
by the Platonic beauty of the sphere—"The sphere," he
wrote, echoing Plato, "is the most perfect . . . the most
capacious of figures . . . wherein neither beginning nor
end can be found"—and he assumed, accordingly, that
the planets move in circular orbits at constant velocities.[1]

Copernicus is celebrated for challenging the grand
assumption of geocentricity. For that challenge he created a scientific revolution. Yet there were more
assumptions that needed to be challenged. As Ferris
notes, Copernicus *assumed* both circular orbits and
constant velocities.

It belonged to Johannes Kepler to challenge and correct these persistent assumptions. Kepler's stormy and
dysfunctional apprenticeship under the martinet
Tycho Brahe is documented by Ferris. Wrestling with
the excruciating problem of the orbit of Mars, Kepler
labored for years with no success. Finally he solved
the conundrum: "My first mistake was in having
assumed that the orbit on which planets move is a circle. This mistake showed itself to be all the more bane-
ful in that it had been supported by the authority of all
the philosophers, and especially as it was quite accept-
able metaphysically."[2]

Ferris relates the conclusion of the tale:

In all, Kepler tested *seventy* circular orbits against
[Brahe's] Mars data, all to no avail. At one point, per-

Johannes Kepler
(1571–1630)

forming a leap of the imagination like Leonardo's to the moon, he imagined himself on Mars, and sought to reconstruct the path the *earth's* motion would trace out across the skies of a Martian observatory; this effort consumed nine hundred pages of calculations, but still failed to solve the major problem. He tried imagining what the motion of Mars would look like from the sun. At last, his calculations yielded up their result: "I have the answer," Kepler wrote to his friend the astronomer David Fabricius. ". . . The orbit of the planet is a perfect ellipse."

Now everything worked. Kepler had arrived at a fully realized Copernican system, focused on the sun and unencumbered by epicycles or crystalline spheres.

Fig. 7

Kepler's Model
of the Universe

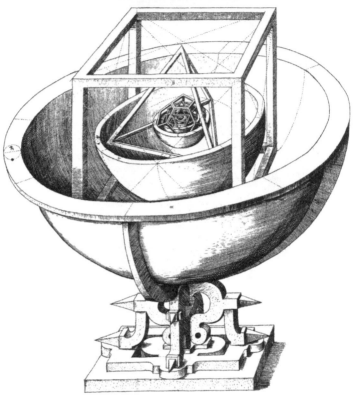

(In retrospect one could see that Ptolemy's eccentrics had been but attempts to make circles behave like ellipses.)[3]

Kepler's laws of planetary motion were a direct result of his challenging assumptions. Once the erroneous assumptions were detected and subsequently corrected, a great leap forward occurred in astronomy.

Fig. 8

Kepler's Diagram
of Planetary Orbits

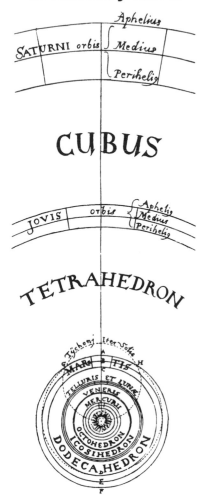

How were the wrong assumptions about planetary motion made in the first place? What led science into a paralysis that would last for centuries? Wrong assumptions were made because *incorrect inferences were drawn from the data.* It was a *deductive* error in the first place.

It is not that earlier scientists made no inductive errors, but the chief culprit was a failure in deduction. Illegitimate inferences led to invalid assumptions, which in turn led to paralysis in progress.

New data gleaned via induction helped the corrective process. The more data we collect, the more frequently we tend to discover anomalies. The more anomalies we encounter, the more pressure that is put on current paradigms. This, in turn, puts pressure on the assumptions until the invalid assumptions are corrected and the paradigm is adjusted.

Faulty inference, then, can function as the archenemy of progress because of its rotten fruit of false assumptions.

Faulty Inference about Chance

The problem of faulty inference with respect to chance may be seen in the illustration that follows.

I heard a minister deliver a sermon in church. He was trying to prove that God is the Creator of the universe. He had read somewhere that the odds against chance creation are astronomical. He gave the calculated odds, which I cannot recall. They were something like one to the jillionth power. The minister concluded from the astronomical probability quotient that it is "mathematically impossible" for the universe to have been created by chance.

When I greeted the minister at the door after the service, he asked me what I thought of his argument. I gently replied that I had three serious problems with it. The first was that if there is one in a jillion chances that the universe was created by chance, then his inference that it is mathematically impossible is unwarranted. If there is one chance in a jillion, it is mathematically possible.

The second objection involves the time factor. If a certain number on a roulette wheel comes up only once every jillion spins, it is possible that this number will come up on the first spin. If the wheel can potentially be spun an infinite number of times, then I presume the odds increase infinitely that sooner or later the number will come up.

My chief objection to the pastor's argument, however, was the third one: If *by* chance means that chance exercised instrumental causal power in creation, then the odds are zero. Again the chances of chance doing anything are nil.

Here the application of logic exposes the false assumptions and illegitimate inferences of the stated argument.

Being contentless, logic cannot, in itself, tell us how the universe began. It can tell us how the universe did not begin.

Challenging Logic Itself

That is true if we assume the validity of logic. But what if this assumption is the big faulty assumption of all time? Ferris said of Kepler that his "penchant for Platonic ecstasy was wedded to an acid skepticism about the validity of all theories, his own included. He mocked no thinker more than himself, tested no ideas more rigorously than his own."[4]

I have been contending for the rigorous application of the laws of logic to inferences drawn from induction. Indeed that is what this book is all about. It is a vehement protest against nonsense statements in scientific theories. Such statements are judged to be nonsense because they violate the laws of logic.

But, we hear, is not logic itself an assumption? Maybe this is the chief assumption that must be challenged in order to pave the way for further progress in science. The policeman with the big stick is just a common bully. If we slay him, we can progress unfettered or unencumbered by his stick.

This is precisely what Niels Bohr did with his assertion "A great truth is a truth of which the contrary is also a truth." Bohr did not shrink from challenging the assumption of logic, thereby sawing off one leg (the deductive one) from the scientist's chair.

We have labored to show that faulty assumptions lead to erroneous conclusions and scientific paralysis that can last for centuries. Such errors are difficult to admit. We have also endeavored to show that though *some* assumptions may be false it does not follow that *all* assumptions are false. We must be careful to distinguish between valid assumptions and invalid assumptions, lest in willy-nilly fashion we throw out the baby with the bath water.

Bohr's challenge is willy-nilly. He challenges the assumption that is the sine qua non of science. To show that logic involves assumption is not difficult. The question is, Is it a *necessary* assumption? To be sure, the assumption is capable of being challenged and even of being denied. But to challenge logic one must first assume its validity in order to make the challenge. To deny logic, one must affirm it.

To say anything intelligible, positive or negative, about logic requires that the words we use in the assertion or denial have intelligible meaning. If the words we use can mean what they mean and their contrary, then they mean nothing and our words are unintelligible.

Philosopher Ronald H. Nash writes:

> Strictly speaking, the law of noncontradiction cannot be proved. The reason is simple. Any argument offered as proof for the law of noncontradiction would of necessity have to assume the law as part of the proof. Hence, any direct proof of the law would end up being circular. It would beg the question.[5]

Nash is correct in his analysis. But again we remember that any attempt to refute the law of noncontradiction also requires one to assume the law being refuted.

When I declare that the law of noncontradiction is a necessary assumption, I mean that without it all other assumptions about anything become impossible. To challenge this assumption makes science an exercise in absurdity. Again, the scientific method itself must be discarded.

People do challenge and deny the law of noncontradiction, but they do so selectively. They deny it when it suits them. Usually this occurs when someone is caught in a blatant logical error. To avoid embarrassment the person says, "I know what I say is contradictory, but that's okay because truth is contradictory." It's a fool's errand to even argue with such persons, because there is no way to test the validity of the argument. The policeman is dead and truth lies slain alongside him.

People who espouse irrationalism, however, tend to argue vociferously for their position. In such cases I refuse to argue because no rules for debate are possible. If they are willing to take a position that they declare up front is absurd, there is no point in my laboring the obvious or in my trying to persuade them that their position is absurd. (Although most people I've met who argue against logic will readily admit to or even glory in the irrationality of it, they will protest if you label their position *absurd*. They want irrationality without absurdity—a difficult request to fulfill.)

Cosmos
or Chaos?

*"A Designer is a natural, appealing and
altogether human explanation of the
biological world. But . . . there is another
way, equally appealing, equally human,
and far more compelling: natural
selection. . . ."*

•

Carl Sagan

The role of mathematics in the progress of science is difficult to overstate. It has functioned through the ages as the avant-garde of scientific breakthroughs and achievement.

Above the door of Plato's Academy in the suburbs of Athens was inscribed the maxim "Let none but geometers enter here." Plato's philosophical work was concentrated in the realm of ideas—of *forms*. His preoccupation with geometry was a preoccupation with the formal and the abstract. Since he considered the empirical world to be the realm of the *receptacle,* the imperfect copies of the archetypical ideas or forms, he focused on what he considered to be the ideal, which to him was also ultimately the real.

It is confusing to students not familiar with technical philosophical language to hear Plato sometimes described as a *realist,* and other times as an *idealist.* The two terms seem to be mutually exclusive. In modern jargon they do refer to disparate concepts. The realist in art or popular philosophy is one who eschews the idealization of things. He paints his heroes' portraits with warts and all.

Plato is called an idealist because he assigns ideas to the highest level of reality. He is called a realist because he attributes *ontological status* to ideas. In his philosophy ideas are not merely names *(nomina),* but also real entities or "things" *(res).*

For Plato, to master the formal is to master the real and therefore to save the phenomena. This may be called a kind of formalism.

Mathematics and Science

Mathematics is a *formal* science. Sometimes the formal dimension of math is understood nominalistically, while at other times it is considered realistically. J. P. Moreland notes:

> Science often uses mathematical language to describe the world. Now, such a use of mathematical language seems to presuppose that mathematical language is true, and if mathematical language is true, there must be mathematical entities called numbers that mathematical language refers to and truly describes.[1]

When Moreland speaks of mathematical language being "true," he is speaking in a realistic way, namely that mathematical numbers are entities that correspond to reality. For mathematical language to be "true," however, it need not represent concrete reality. Statements may be judged *formally* "true" even if they have no correspondence to *material* reality. We may use mathematical language to describe or quantify material reality, but the language is still formal. Moreland continues:

In sum, since science often uses mathematical language in a way that assumes the truth of mathematics, and if the truth assumed in mathematics is to be understood in the same way that truth is involved in fields outside mathematics, then the truth of mathematics carries with it an ontological commitment to abstract, mathematical entities that ground the truth of mathematical claims. Thus, some argue, science presupposes the truth of mathematics and, therefore, the existence of abstract, mathematical entities.[2]

The question Moreland is raising with respect to the "truth" of mathematics is the question of the relationship of the mathematical to the ontologically real. It is the question of the relationship of the formal to the material.

The classic issue of this relationship may be seen in the struggle between seventeenth-century rationalism and eighteenth-century empiricism. It was not by accident that seventeenth-century rationalism was dominated by mathematicians, such as René Descartes and Baruch Spinoza. The world had been revolutionized by the spectacular success of applied mathematics. Nicolaus Copernicus's revolution was spawned by a new mathematical paradigm. Galileo's discoveries in the heavens were prompted by mathematical formulas that told him where to point his telescope.

Later the age of Newtonian physics was heralded by mathematical discoveries, and Albert Einstein, the math wizard, fathered the advent of the atomic age.

The question raised in the seventeenth century revisited Plato's Academy. Conceptualism arose, arguing that if something is true formally it must also be true materially.

The idea that the formal must indicate or be recapitulated by the material is not the same as the idea that the material can be described by the formal. To state it another way, the question is this: Is the logical necessarily the real, or is the real logical? Can the formally true be materially false? Can the materially true be formally false?

These questions probe the heart of the matter. The ultimate question is this: Is reality rational or irrational? Is the universe cosmos or chaos?

How did we suddenly get back to the question of logic from a brief survey of the role of mathematics in science? The movement is not a quantum leap. The connection appears simple and obvious if we make one observation: *Mathematics is a form of symbolic logic.*

If logic is the science of the rational relationship of propositions, then we see the connection. Mathematics is abstract. It uses ciphers to represent concepts. A number is a symbolic representation. Mathematics is an abstract form of speech similar to symbolic logic. When we say $2 + 2 = 4$, we are using a formal, logical equation. When we apply this to material reality, we may say 2 apples + 2 apples = 4 apples. Whether we are describing a quantity of apples, oranges, or steamboats, the formal relationship remains the same in the equation.

There is no essential difference between the mathematically "true" and the logically "true." Likewise the

Galileo Galilei
(1564–1642)

mathematically false and the logically false are inter-
related.

Math is the logic of the universe. This is why math
serves as the strongest arm of the deductive side of the
scientific method.

Just as errors can be made in logic, so errors can be
made in math. We've all heard the attempts to enlist
mathematics in the cause for an irrational view of
physics in reality. We've heard the argument that
$2 + 2 = 5$, or that $2 = 1$. William Poundstone analyzes
the latter:

> The weakest type of paradox is the fallacy. This is a
> contradiction that arises through a trivial but well-
> camouflaged mistake in reasoning. We've all seen

those algebraic "proofs" that 2 equals 1, or some other absurdity. Most are based on tricking you into dividing by 0. One example:

1. Let $x = 1$
2. Then obviously: $x = x$
3. Square both sides: $x^2 = x^2$
4. Subtract x^2 from both
    sides: $x^2 - x^2 = x^2 - x^2$
5. Factor both sides: $x(x - x) = (x + x)(x - x)$
6. Factor out the $(x - x)$: $x = (x + x)$
7. Or: $x = 2x$
8. And since $x = 1$: $1 = 2$

The fatal step is dividing by $(x - x)$, which is 0. Line 5, $x(x - x) = (x + x)(x - x)$, correctly asserts that 1 times 0 equals 2 times 0. It does not then follow that 1 equals 2; any number at all times 0 equals any other number times 0.[3]

We notice that in this analysis Poundstone speaks of "a contradiction that arises through a trivial but well-camouflaged mistake in reasoning." He speaks of a *logical* mistake camouflaged in a *mathematical* formulation.

The Mathematical and the Real

Let us return now to our question about the relationship of mathematics to the real world. In essence it is the same question as the relationship of the logical or rational to the real world.

The objection eighteenth-century empiricism raised against conceptualism is that, though everything that is real is rational, it does not follow that everything that is rational is necessarily real. Ideas can be conceived that are perfectly rational but that have no corresponding empirical reality.

In analyzing the function of the mind, John Locke, who argued that at birth the mind is a *tabula rasa*, assigned certain abilities to the mind to process both data and ideas drawn from sensory data. He said that the mind has the ability to *combine*, *abstract*, and *relate* ideas. This is what makes fiction, fantasy, and creative

John Locke
(1632–1704)

From engraving by Pofselwhite

imagination possible. I can conceive of an animal called a unicorn by abstracting certain body parts from known, real-life animals and combining these ideas in a new way. We can "create" new ideas, like the idea of a unicorn, but it does not mean that we can actually create real, immaterial unicorns.

This is the stuff that horror movies and science fiction are made of. We respond emotionally to the plight of E.T. as we behold his image on the silver screen. Our children sob when he dies; we lean over to reassure them that it isn't real, it's only pretend.

It is because we can have abstract ideas that are rationally possible that imagination is so much fun. Yet we must remind ourselves that, though a concept may be rationally or logically *possible,* this does not mean it is real.

On the other hand if the real cannot be rationally impossible, then logic as the governor can serve us well in our search for the truth about reality. If the real is rational, then when our concepts are demonstrated to be irrational, we know our concepts are out of touch with reality.

How do we know that the real is rational? We don't. What we do know is that if it isn't rational, we have no possible way of knowing anything about reality.

That the real is rational is an assumption. It is the classical assumption of science. Again it is a *necessary* assumption for science to be possible.

If the assumption is valid and reality is rational and intelligible, then the falsifying power of logic can play a major role in scientific inquiry.

God and Logic

When metaphysicians and theologians describe God as a self-existent, eternal being, they are using a *concept* that is at least logically *possible*. There is nothing illogical about the concept of self-existent being. The idea commits no formal error.

Now because the concept is possible logically, it does not necessarily follow that there must be a self-existent being in reality (at least not yet).

On the other hand, as we have labored to show, the *concept* of *self-creation* is logically impossible. If in arguing about the nature of ultimate reality we were able to reduce the possible options to two, self-existence or self-creation, we would be driven to embrace the former and reject the latter. To select the former (self-existence) is to select the logically possible. To select the latter (self-creation) is to select the logically impossible.

But the trick here is to reduce the options to the logically possible and the logically impossible. When considering the question of the origin of the cosmos, we have several options (see fig. 9): (1) the cosmos is an illusion—it doesn't exist; (2) the cosmos is self-existent (and eternal); (3) the cosmos is self-created; (4) the cosmos is created by something that is self-existent.

Are there other options I've overlooked? I've puzzled over this for decades and sought the counsel of philosophers, theologians, and scientists, and I have been unable to locate any other theoretical options that cannot be subsumed under these four options. We've already seen, for example, that spontaneous generation or chance creation are, under analysis, the same as option 3 (self-creation).

The most frequently mentioned "fifth option" is Bertrand Russell's infinite series of finite causes, or the so-called infinite regress. That option is, however, simply a thinly disguised or camouflaged form of option 3. It is self-creation camouflaged to infinity.

If we look at the four options above, we can quickly eliminate two of them. Option 3 has to go because it is formally false. It is contradictory and logically impossible.

Option 1 can also be safely eliminated, for two reasons. First, if the cosmos is an illusion, then we still have to account for the illusion. If it is a false illusion, then it isn't an illusion. If it is a "true" illusion, then something or someone must exist to have the illusion. If so, then whatever is having the illusion must be self-created, self-existent, or created (caused) by something that is (ultimately) self-existent.

The second reason we can eliminate option 1 is that, if the illusion is absolute in the sense that nothing exists, including whatever is having the illusion, then the jig is up. We have no question to answer because nothing exists.

However, *if* something exists, then whatever exists is either ultimately self-existent or created by something that is self-existent.

We notice that if the above is true, then in either case we encounter something that is self-existent. Now we

Fig. 9

Origin of the Cosmos: Four Options

Option 1   The cosmos is an illusion; it doesn't exist.

Option 2   The cosmos is self-existent (and eternal).

Option 3   The cosmos is self-created.

Option 4   The cosmos is created by something that is self-existent.

begin to see that the concept of self-existent reality is not only logically *possible,* but also logically *necessary.*

In shorthand form the argument is this: If something exists, then something, somewhere, somehow, is self-existent. The only option to this is self-creation, which is logically impossible. This is a modern restatement of St. Thomas Aquinas's argument of *necessary being* (*ens necessarium*). Aquinas argued that the existence of God is both *ontologically* and *logically* necessary.

Self-existent being is being that has the power of being within itself. It is not dependent on anything outside itself in order to be. Since self-existent being is independent being, it has no beginning. It always has been. It is remarkably similar to Parmenides' "Whatever is, is." It cannot not be.

To put the issue another way. If there ever was a "time" when nothing existed, what would exist now? Obviously nothing (unless something can come from nothing, which puts us back to self-creation). So we know logically that *if* something exists now, then *some-thing* is self-existent. Self-existence is now a logically necessary concept.

If *something* must be self-existent, then the question is, What (or who) is it? Many, if not most, thinkers who deny the existence of God readily admit that logic demands that something be self-existent. They disembark the theologians' boat when the theologians insist that the self-existent "something" must be God.

Is the self-existent being God (the Creator), or is the self-existent being the cosmos? The question can be broken down further. If we say that the cosmos is self-existent, do we mean that each *part* of the cosmos is self-existent, or that some part of the cosmos is self-existent and produces, generates, or causes (creates) the other parts? If all parts of the cosmos are self-existent, then I am self-existent and my watch is self-existent. Yet I know that I am not self-existent and neither is my watch. We were brought into existence. There was a time when we were not. We are contingent, dependent, derived entities.

If we argue that it is not necessary to argue for a *transcendent* self-existent being but locate the power of being or self-existence *within* the universe, we make a fatal conceptual error. We reveal a serious problem of linguistic confusion with respect to the meaning of the word *transcendent*. When theologians say that God is transcendent, they are not speaking in spatial or geographical terms. They are not describing where God lives. Transcendence refers to God's *ontological* status with respect to the world. God, by virtue of self-existence, is a higher order of being than that which is not self-existent.

If we argue that one part of the universe is self-existent and has the power of being within itself by which it can generate lesser levels of existent reality, then we have attributed to this mysterious being-within-the-universe the attributes of a transcendent God. A rose by any other name . . .

Another issue regarding self-existence is encountered whenever we speak of other universes having a beginning. Self-existent beings do not "begin" to be. If they begin to be, they are not self-existent.

Is the point of singularity, which is thought to have exploded fifteen to seventeen billion years ago, the self-existent being we are looking for? Perhaps. But if this highly condensed piece of matter and energy preexisted the Big Bang for eternity, what caused it to explode? Did self-existent being suddenly disintegrate? Was this eternally inert "piece" of reality acted upon by an outside force, or did it defy the laws of inertia?

However we answer these questions, we are still left with the logically necessary idea of self-existence. Logic cannot examine the phenomena. It can and does govern and monitor our inferences drawn from the phenomena, and it alerts us when we stray into the realm of the logically impossible.

Linguistic Confusion Revisited

Logic functions as a policeman not only when we indulge in irrational concepts like self-creation. We find other oxymorons popping up from time to time in discussions of science (and in philosophy, theology, and every other field of inquiry).

Linguistic confusion occurs when analytically false statements are used or when "studied ambiguity" replaces linguistic precision. Precision in speech is an important complement to precision in research.

When phrases such as "inherent randomness" or "random selection" are used, we wonder what they mean.

If a particle is said to have "inherent randomness," this suggests that a random character is built in or intrinsic to it. Does this mean that it acts the way it does for no reason at all? Or does it merely mean that its activity cannot presently be predicted? The latter is a posture of humility commensurate with the real limitations of our knowledge. The former suggests an illogical concept, namely that the activity of the particle is an effect without a cause.

The phrase *random selection* (or *random mutation*) is also somewhat confusing.

To make a "selection" suggests some sort of *intentionality,* a trait usually associated with *intelligence.* Is it possible to have an unintentional intention?

In popular jargon we use the phrase *random selection* to describe certain types of actions. Suppose we are going to draw a ticket stub from a box in order to award a door prize to a "lucky" participant. We shake the box to insure a proper mixing and perhaps even blindfold the person designated to select the ticket. We want to choose the ticket "at random" to insure that the drawing is not rigged in favor of a particular contestant. We seek to leave the outcome to "chance."

This is a legitimate use of the phrase *random selection*. A selection is made but without the specific intention of choosing a particular person's ticket. But intention is at work. We are *intentionally* choosing an unknown ticket.

Again "chance" does not influence which ticket is chosen. That is determined by how the hidden tickets were mixed in the shaking and where the selector's hand reaches into the box. There is no operation of chance itself.

If the phrase *random selection* is used as a synonym for action-without-a-cause, then it is illogical.

Also the term *selection* may be used in a metaphorical or figurative sense. This is the way the phrase *natural selection* is often used.

Carl Sagan muses: "A Designer is a natural, appealing and altogether human explanation of the biological world. But, as [Charles] Darwin and [Alfred] Wallace showed, there is another way, equally appealing, equally human, and far more compelling: natural selection, which makes the music of life more beautiful as the aeons pass."[4]

Sagan goes on to speak of "fate," "accident," and "chance" as factors involved in natural selection. These words seem to militate against the idea of actual *selection.* If selection involves intentionality, then the concepts collide. We cannot have an intentional accident or an accidental intention.

Sagan, however, does not seem to suggest that natural "selection" involves *intentionality.* A snowflake doesn't "intend" to land where it lands. It is subject to wind currents and the like. The snowflake lands where it lands for a reason. Its flight has a cause. Its selection of a landing site is not intentional and as such is a "selection" only in the figurative sense.

I have no quarrel with the use of metaphorical or figurative language in describing natural phenomena, as long as we understand that is what we are doing.

The confusion arises, as we have seen, when we push the language to a literal sense by attributing causal power to "chance" and "randomness."

The difficulty we face in this process is gaining a clear understanding of the nature of causality. Since this question is so central to understanding the behavior of phenomena, we will devote an entire section of this work to it.

A Being
without
a Cause

"If everything must have a cause, then God must have a cause. If there can be anything without a cause, it may just as well be the world as God. . . ."

•

Bertrand Russell

The concept of causality is one of those ideas that enjoys constant usage in both popular language and technical jargon. *Cause* is a loaded term suffering from the abuse of casual and inexact usage. Errors in usage complicate attempts to understand the inner workings and connections among phenomena.

In popular or ordinary language the concept of cause is used to cover or explain a multitude of situations. When a child asks why the rain falls, the simple reply he often hears is, "Because."

The term *because* is used to explain something. It seeks to provide a reason for things as they are. It assigns power or locates influence. It seeks to answer the "why" questions of life. It has been argued that any person's knowledge can be exhausted by seven questions, as long as they begin with the question "Why?" and are asked seriatim.

Only moments ago, as I finished writing the above statement, a bone surgeon walked into the room. I decided to test the seven-question theory on him. I asked him, "Why do bones break?" He answered it promptly. By the fifth successive "why" question, he shrugged his shoulders and said, "I don't know."

If we ask why long enough, sooner or later we get to the question "Why is there something rather than nothing?" The answer to that question is usually "Because."

A Hypothesis Concerning Causality

The question of causality begins with the problem of definition. A frequent formulation of the law of causality is the maxim "Everything has a cause." If we analyze this proposition, we discover a certain ambiguity lurking under the surface. When we say everything has a cause, do we mean simply that there is a *reason* for everything that is? Or do we mean that behind every thing that is, there is something else antecedent to it?

If we mean the latter, namely that every thing that is has an antecedent cause, we have consciously or unconsciously stepped into the abyss of absurdity by which we must embrace an infinite series of antecedent causes. The problem of such a series is that it can never get started. We must either begin the series with a cause that has no antecedent or postulate an infinite series which series has no antecedent. It compounds the problem of self-creation infinitely.

The maxim "Everything has a cause" in the above antecedent sense is fatal to theology. If God has an antecedent cause, then God is not God. If God exists but has no antecedent cause, then everything does not have an antecedent cause.

Some seek to get around this problem by asserting that God, or the first cause (whatever it is), is self-caused. It is similar to the discussion between two children when one asks the other, "Where do trees come from?" The second replies, "God made the trees." The first counters, "Who made God?" The reply is, "God made himself."

The problem, of course, is that even God, no matter how powerful he may be, cannot make himself. For God to make himself he would have to be before he was. In other words, he would have to be and not be at the same time and in the same relationship, which violates the law of noncontradiction and necessitates a leap into the absurd.

To say that God causes himself suffers from the same fatal error that afflicts any proposition about something causing itself: It is nonsense. It reduces to the concept of something being created, made, or caused by nothing.

Traditional philosophy argued for the existence of God on the foundation of the law of causality. The cosmological argument went from the presence of a cosmos back to a creator of the cosmos. It sought a rational answer to the question "*Why* is there something rather than nothing?" It sought a sufficient reason for a real world.

We will return to the cosmological argument later. For now we merely note in passing that it has enjoyed a long and tenacious history in philosophical reflection, especially as it was articulated by Thomas Aquinas.

The cosmological argument for God has been widely dismissed in the modern era, chiefly for three reasons. The first is the comprehensive criticisms leveled against it by Immanuel Kant. The second is the assumption that the very principle of causality on which it rests was demolished by David Hume. The third is that quantum physics proves that something can come from nothing.

It is the hypothesis of this writer that all three reasons are spurious. In this section we will devote considerable space to the considerations of David Hume. Before we do that, however, we must press on with the question of initial definitions.

The statement "Everything has an antecedent cause" contains within it a definition of causality that is precisely what is in dispute. If this is our definition of causality, then we have a hypothesis, not a law. If it functions as law, it does so gratuitously and, as I shall endeavor to show, irrationally. The law as stated can neither be demonstrated nor verified empirically, nor can it be verified by rational analysis.

It cannot be verified empirically because it makes a universal affirmation about everything. To verify such an ambitious law would require that we know something, at least, about everything that is. The something we would have to know is that it has an antecedent cause. How could we ever know that we know everything that is? We would not only need to know everything that is, but we would have to know that everything that is has an antecedent cause. Yet to know that would be to know that we know nothing, because if everything has an antecedent cause, reality must be irrational. If reality is irrational, we can know nothing because rationality is a necessary condition for knowledge.

I assert, then, that to define the "law" of causality by saying that "everything has an antecedent cause" is utterly arbitrary and not worthy of science. It is a lousy law, a law decreed by arrogant fiat.

The Law of Causality

The "law of causality" is usually linked with the word *effect*. This law is often called the "law of cause and effect." Now we are getting to an axiom that makes more sense.

We usually define *effect* as that which has an antecedent cause. Cause and effect, though distinct ideas, are inseparably bound together in rational discourse. It is meaningless to say that something is a *cause* if it yields no *effect*. It is likewise meaningless to say that something is an *effect* if it has no *cause*. A cause, by definition, must have an effect, or it is not a cause. An effect, by definition, must have a cause, or it is not an effect.

Another way of defining the law of causality is this: *"Every effect has an antecedent cause."* This definition has a huge advantage over the definition *"Everything has an antecedent cause."* The difference is between every *thing* and every *effect*. We do not know that every thing has an antecedent cause. We do know that *every effect* has an antecedent cause.

The advantage of the former definition is that it is formally valid. The statement "Every effect has an antecedent cause" is *analytically true.* To say that it is analytically or formally true is to say that it is true by definition or analysis. There is nothing in the predicate that is not already contained by resistless logic in the subject. It is like the statement "A bachelor is an unmarried man" or "A triangle has three sides" or "Two plus two are four."

As a formal definition it tells us little or nothing about reality. The law of causality, so defined, is merely a logical extension of the law of noncontradiction. The law doesn't tell us if there really are such things as causes or if there really are such things as effects. What it does tell us is that *if* there are causes those causes must have effects, and that *if* there are effects those effects must have causes.

Logic requires that if something exists contingently, it must have a cause. That is merely to say, if it is an effect it must have an antecedent cause. Logic does not require that if something exists, it must exist contingently or it must be an effect. Logic has no quarrel with the idea of self-existent reality. It is logically possible for something to exist without an antecedent cause. It remains to be seen if it is logically necessary for something to exist without an antecedent cause. For now it is sufficient to see that self-existence is a logical possibility. The idea is rationally justified in the limited sense that it is not rationally falsified. Something is rationally falsified when it is shown to be formally or logically impossible.

We have already labored the point that the idea of self-creation is formally falsified. It fails the test of logical analysis. Again, for something to create itself it must defy the law of noncontradiction to do it. No such burden attends the concept of self-existence.

Logic does not require that there be effects in reality. Logic does not require that there be causes in reality. Bare logic does not require that there be a self-existent something. Indeed bare logic does not require that there be any kind of something. Logic is indifferent to the question of whether there is something or nothing at all, with the exception that there is the logical impossibility of there *being* nothing, because nothing, by definition, does not exist. We cannot even conceive of nothing because to conceive of it would be to conceive of something, and something is not nothing.

The First Cause

The fact that the idea of self-creation is rationally falsified and has no formal justification does not mean no one will ever affirm it. The idea of self-creation is *sayable;* it is not rationally conceivable. I can say that triangles have four sides, but I cannot conceive of one. In the famous debate between Bertrand Russell and Frederick Copleston on the existence of God, Russell kept insisting that the universe could be the product of an infinite series of finite causes. Copleston insisted that such a notion is rationally inconceivable, though he granted that Russell could *say* it.

Russell personified the modern era's antipathy to causal reasoning. In his celebrated essay "Why I Am Not a Christian," he writes about arguments the church has used in its attempt to prove the existence of God.

Perhaps the simplest and easiest to understand is the argument of the First Cause. (It is maintained that everything we see in this world has a cause, and as you go back in the chain of causes further and further you must come to a First Cause, and to that First Cause you give the name of God.) That argument, I suppose, does not carry very much weight nowadays, because, in the first place, cause is not quite what it used to be. The philosophers and the men of science have got going on cause, and it has not anything like the vitality it used to have; but, apart from that, you can see that the argument that there must be a First Cause is one that cannot have any validity. I may say that when I was a young man and was debating these questions very seriously in my mind, I for a long time accepted the argument of the First Cause, until one day, at the age of 18, I read John Stuart Mill's Autobiography, and I there found this sentence: "My father taught me that the question, 'Who made me?' cannot be answered, since it immediately suggests the further question, 'Who made God?'" That very simple sentence showed me, as I still think, the fallacy in the argument of the First Cause. If everything must have a cause, then God must have a cause. If there can be anything without a cause, it may just as

Bertrand Russell
(1872–1970)

well be the world as God, so that there cannot be any
validity in that argument.[1]

This lengthy quote reveals a fascinating pilgrimage
in Russell's thought. He indicates that in his youth he

was impressed by the First Cause argument. In his adolescence he was impressed by an argument from a formidable philosopher, John Stuart Mill, and this argument remained convincing to him for his entire life.

Let us evaluate Russell's comments. He contends that the First Cause argument no longer carries much weight "because" [sic!] cause is not what it used to be. Here we note that, though in his judgment cause is not what it used to be, it is not so antiquated that Russell himself will not appeal to it. His "because" is an attempt to assign a cause for the decline of causal reasoning. He then notes the cause of the decline of cause: "The philosophers and the men of science have got going on cause. . . ."

It is true that certain philosophers have indeed "got going" on cause. We will examine this later.

Russell then asserts that "the argument that there must be a First Cause is one that cannot have any validity."

If we take this proposition as a serious and sober assertion and not as an ejaculation of hyperbole, we see that Russell is making a strong statement. He says the argument *cannot* have *any validity.* This proposition is a universal negative. It says that the argument is *not able* to have *any* validity. It is not possible for the First Cause argument to have the slightest hint of validity.

Russell is not satisfied to make a mere bold or dogmatic assertion about this: he then attempts to show *why* the First Cause argument cannot possibly have any validity. Leaning on Mill's anecdote he pinpoints what he calls the "fallacy" of the First Cause argument:

If everything must have a cause,
then God must have a cause.

His argument is valid. It fits the classic *modus ponendi* form. Here we see a classic argument that is *valid* but its conclusion is not sound. The conclusion follows resistlessly from the major premise. The problem with it rests, however, in that major premise: "*Everything must* have a cause." How can we rationally justify this assertion? What law teaches that everything that is, is necessarily an effect? We have seen already that the concept of self-existent being is rationally possible. It violates no law of reason. If that is the case, we can offer another argument:

If something may be self-existent,
then something may exist without a cause.

What is astonishing about Russell's argument is that in the same paragraph he writes: "There is no reason why the world could not have come into being without a cause. . . ."[2] Here Russell does two significant things. First, he demonstrates the tenuous character of the major premise of the original argument. He grants explicitly that something (namely, the world) may exist without a cause.

There is a subtle point of reasoning here that must not be overlooked. That Russell allows for something to exist without a cause does not vitiate or lessen the force of his previous argument. His previous argument, a *modus ponendi* argument, remains valid. We remember that the argument is *conditional:* "*If* everything has a cause, then God must have a cause." He is saying here, not that everything in fact has a cause, only that if we argue that everything must have an (antecedent) cause, then God must also have a cause.

We have no dispute with this. The problem, as we've already noted, is that the argument assumes a specious form of the law of causality. If we assume that the law of causality teaches that every *thing*, not merely every *effect*, must have a cause, then Russell is right.

What this reveals, however, is that both Russell and Mill, in attacking the First Cause argument for God, attack a straw man. If the First Cause argument rested on the premise that everything must have a cause, then Russell's judgment of its validity would be absolutely correct. It could not possibly have *any* validity.

Perhaps it should be unnecessary to say that the First Cause argument does not rest on this faulty premise. However, since men of the prodigious intelligence of Russell and Mill missed the crucial distinction between "Every effect must have a cause" and "Every thing must have a cause," it is necessary to stress it once more.

The second step Russell takes is fatal. In it he makes two different assertions: (1) there is no reason why the world could not have come into being without a cause and (2) there is no reason why the world could not have always existed.

These two assertions differ, and they differ profoundly. The first assertion is simply incorrect. The world could not have come into being without a cause, for this would have required some sort of self-creation. We have insisted *ad nauseum* that the concept of self-creation is falsified by the law of noncontradiction. There is every reason why the world cannot come into being without a cause. Reason itself is the reason why this premise must be rejected. It asserts a formal impossibility. What Russell deems possible for the world to do—come into being without a cause—is something no judicious philosopher would grant that even God could do. It is as formally and rationally impossible for God to come into being without a cause as it is for the world to do so.

The second assertion is not so irrational. Here Russell postulates the possible existence of an eternal world. The idea of a self-existent, eternal world is not formally false. That is, it is not a rationally impossible concept.

Indeed, reason demands that if something exists, either the world or God (or anything else), then *something* must be self-existent. The only possible alternative to this is that something comes into being without a cause or is self-created.

If Russell's world comes into being without a cause, it is neither an effect nor a cause of itself. Since it has no cause, it cannot be an effect. Indeed if it comes into being without any cause, it cannot even be its own cause. Apart from the absurdity of saying something is its own cause, Russell raises the absurdity to a higher level by assigning no cause whatever.

Russell is guilty here of sloppy speech, if not confused thinking. He probably means that the world comes into being without an antecedent cause or some external cause. He is probably guilty of the lesser absurdity (if absurdity can be measured in degrees) of assuming some sort of self-creation.

Let us suppose that the absurdity of self-creation is possible (a really impossible supposition). For something to bring itself into being it must have the power of being within itself. It must at least have enough causal power to cause its own being. If it derives its being from some other source, then it clearly would not be either self-existent or self-created. It would be, plainly and simply, an effect.

Of course the problem is complicated by the other necessity we've labored so painstakingly to establish: It would have to have the causal power of being before it was. It would have to have the power of being before it had any being with which to exercise that power.

A Self-Existent Entity

The purpose of this exercise in absurdity is to get relentlessly to the bottom line. Something must have the power of being within itself (*self-existence* or *aseity*) or nothing would or could possibly exist. This is the heart of the First Cause argument that Russell embraced in his infancy but discarded in his adolescent years. This is the point that apparently eluded Russell for his entire life, leaving him in a logical quagmire.

The force of the First Cause argument is this: *If* something exists, something somehow, somewhere, at some time has the power of being intrinsically. It is not an effect. The only possible logical alternative to a First Cause is No Cause. An infinite series of finite causes merely confuses the issue by raising the problem of self-causation to the infinite level of absurdity.

The rational alternative of *No Cause* is possible only if we eliminate the "if" from the premise "If something exists . . ." If there is no cause for the world, then only two possibilities remain: Either there is no world, or the world is self-existent and eternal.

In his deliberations Russell seems to have difficulty escaping the conclusion that if something exists it must have the power of being within itself. Something must possess self-existence. The only question left to explore is this: What is the self-existent something?

Russell allows for the possibility of a self-existing world that is not an effect. If it is self-existing, then by definition it is not an effect.

Theists argue that there must be a self-existent First Cause and that this cause is God. Russell argues that the First Cause may be the world itself.

This poses the following question: When we say that the world is self-existent, do we mean that everything in the world is self-existent or that only part of the world is self-existent and causes other effects?

If everything in the world is self-existent, including you and me, then, manifestly, there would be no effects whatsoever in the world.

Suppose, however, there are some effects in the world. Suppose the entity with which I am writing, my pen, is an effect—caused by something else. Suppose this pen is contingent, derived, and dependent as an entity. (Perhaps its composite elements self-exist, but the discrete, individuated entity I am using, in its present form, is not self-existent.)

If my pen is not self-existent, then there is something in the world that must be distinguished from whatever part of the world that is self-existent. The question is, In what way must the pen be distinguished from the self-existent part or parts of the world? The distinction in this case must necessarily be one of ontology. The ontological distinction between the pen and self-existent others is that the pen is *not* self-existent. The pen is *caused* by something antecedent to it, the pen-maker or the pen-making machine. The pen's individuated, discreet existence is caused by something or someone other than the pen.

The salient point is this: If some things in the world are self-existent and other things are contingent, then there is a necessary distinction to be made between them ontologically. The self-existent realities *transcend* the contingent realities in an ontological way.

The term *transcendent* can be slippery and confusing. Traditional theological language speaks of the transcendence of God as distinguished from his immanence. Transcendence refers to that sense in which God is "above" or "beyond" the world. Words like *above* or *beyond* are borrowed from spatial or geographical language. The clouds are "above" the ground and Europe is "beyond" the sea from America.

When the term *transcendent* is applied to God, however, it does not refer to God's location or physical stature. It does not mean that God is bigger, fatter, or taller than creatures. Nor does it mean that he lives way up in the sky somewhere east of the moon and west of the sun. The term refers specifically to the order of being God represents. It refers to his ontological status. When theologians say God is a transcendent being, they mean that he transcends every created thing ontologically. He is a higher order of being precisely at the point of his being. The specific point is that he is a self-existent and eternal being who has the power of being in himself. He is uncaused. He is self-existent.

The Universe

The issue of a self-existent, eternal universe presents us with another confusing and slippery concept. It is the concept of the "universe" itself. The term *universe* suffers from frequent equivocation. Sometimes *universe* is used to refer to "everything that is or exists."

If the universe refers to all that exists and *if* God exists, then it is a necessary logical inference that God is a part of or within the scope of the universe. He is a particular within a universal. Again, if the universe is composed of all that exists or is, that is, of *all beings,* then both self-existent being and contingent being are a part of or contained within the scope of this word.

On the other hand the term *universe* is often used elliptically. There is an ellipse, an unspoken element that is tacitly assumed. That is the restricted qualifier *created.* In simple terms the word *universe* is often used and understood to refer to all of creation, excluding God. In this usage God is not considered part of the universe but stands over and above the (created) universe as its transcendent Author or Cause.

The emergence of the word *universe* comes from a mongrelization of two distinct words that are blended into one. The two words are *unity* and *diversity.* Again the assumption is that there is a transcendent reality whose power of being gives impetus and coherency to a wide diversity of contingent things.

When philosophies speak of an eternal universe, what is usually meant is that *some parts* or *part* of the universe is self-existent, not that each distinct, individuated entity is self-existent. That is, there is still room for *effects* or contingent entities. If there are such effects, then there must also be antecedent causes. We also recognize that something can be both a cause and an effect *at the same time,* but *never in the same relationship.* It is possible to have a chain of causes, causes that are themselves effects of previous causes.

If we speak of the universe in these terms and seek to locate something *within* the universe that is self-existent and uncaused, the ultimate source of causal power from which every effect in the chain derives its existence, then we must be careful to define what we mean by *within*. Such a self-existent being may be contained *within the concept* of universe (if we are using the first definition of universe, "all that is"), but it still must be regarded as ontologically transcendent to all that is contingent upon it. It is within the *concept* of the word *universe,* but it is not within the *concept* of the *created universe.*

If we posit some arcane, unknown, pulsating source of the power of being *within* the universe, we are still speaking of something that is *beyond* the rest of the universe in terms of its being. Its *location* is within; its *being* is above and beyond everything else.

An Eternal Being

We see then that we can cut the Gordian knot of transcendence and press to the bottom line: there must be a self-existent being of some sort somewhere, or nothing would or could exist.

A self-existent being is both logically and ontologically necessary. It is in its purest sense *ens necessarium*, "necessary being." We have labored the *logical* necessity of such being. Yet it is also necessary ontologically.

An ontologically necessary being is a being who cannot not be. It is proven by the law of the impossibility of the contrary. A self-existent being, by its very nature, must be eternal. It has no antecedent cause, else it would not be self-existent. It would be contingent.

A self-existent, eternal being exists by its own intrinsic power of being. It is dependent on nothing outside itself for its own being. Since it exists by its own eternal power of being, it is altogether underived, independent, and invulnerable to anything outside itself.

Philosophers may quarrel and ask, Why call this necessary being, which is both logically and ontologically necessary, by the name *God?* Why not call this mysterious, self-existent being Q? Why not call it Super Energy or Cosmic Energy or something else?

The simple and short answer to this question is another question: Why not call it God? From a scientific, metaphysical, or philosophical perspective it doesn't matter what you call it. What matters is the concept or the reality, not the name or word used to indicate it. A rose by any other name is still a rose.

A Higher Being

From a theological perspective it matters profoundly what we call it. The word *God* connotes far more than "self-existent, eternal being" but by no means less than "self-existent, eternal being." The word *God* also indicates a holy and personal being who is worthy of worship and to whom all persons are held ultimately accountable. It is one thing for me to relate to God; it is quite another thing for me to relate to Super Energy or cosmic dust. It is the difference between a relationship between two personal beings (God and me) and a relationship between a personal being (me) and an impersonal being (Super Energy).

We see this problem in the currency of common language about God. It is a popular mode of expression to refer to God as a "higher power." The term *higher power* is a studied ambiguity, a necessary concession to the idea that there must be a source to the power supply of being.

Power in itself, however, is not necessarily personal. We distinguish between the essence of a thing and its power. Power divorced from being is no power—it is nothing. If there is power present, there must be something generating that power.

I once had a discussion with a physics professor from Carnegie Mellon University. I was lecturing on the difficulties inherent in human language about God during the height of the God-talk controversy of the sixties. The professor was somewhat disdainful about the whole business of theology. He did not consider it a science because its language about God is so imprecise. He said to me: "You really don't know what God is. The language is too imprecise and ambiguous." I didn't concede his point but argued in an *ad hominem* fashion. I did grant that language about God is difficult but insisted that it is necessary, helpful, and meaningful. I said, "Surely you can identify with this problem as you have the same problem in your discipline of physics." He challenged me, saying he had no similar difficulty.

I said to him, "Please tell me what energy is."

He said, "That's easy, it is the ability to do work."

I replied, "I didn't ask you what it can do, I asked you what it *is*."

He then answered, "Energy is MC^2."

Again I said, "I don't want to know its mathematical equivalence; I'm asking about its ontology—what *is* it?"

The professor began to feel the weight of the problem. He wanted to restrict the meaning of energy to naked power, and yet he wanted to use the term as if it had being. If energy is simply the power or force exerted by some kind of being, it is a perfectly meaningful term. My objection was aimed at the not-so-subtle equivocation that goes on when *energy* is used to refer to something that exists in itself.

Simply stated, there cannot be any "higher power" unless there is also a *higher being* who generates that power.

That it is customary to refer to God as a higher power rather than a higher being raises an interesting psychological question. Since it is obvious, or at least should be obvious, that there can be no power without something generating it, why are we inclined to substitute power for being?

Since I have written an entire book on this subject, *The Psychology of Atheism*,[3] I will not elaborate on the point here, other than to summarize by saying there is an attractive escape hatch to evade personal moral culpability by reducing God to an amorphous, impersonal power. Cosmic dust or Super Energy does not say, "Thou shalt not steal." I never have to worry about being judged personally by an impersonal force.

Apart from psychological considerations there are other reasons why the term *God* is appropriate for self-existent, eternal being. The chief is that the major point of debate concerning the existence of God focuses on the necessity of believing in a self-existent, eternal Being. Again, though God has other attributes besides eternality and self-existence (aseity), his self-existence is what defines his transcendence.

In philosophical and metaphysical theory it has been customary, historically, to assign the name *God* to the uncaused Being who is the ultimate source of all other beings. This transcendent ultimacy lends credence to the idea that this Being is worthy of our worship. Since we owe our life and all the benefits we receive in the created order to this source of all being, it is fitting that we acknowledge our dependence on him and offer our gratitude in the sacrifice of praise.

A Personal Being

To be sure, this assumes that this self-existent eternal one is a "him" and not an "it." It assumes that this Being is *personal.*

The theological ascription of personality to self-existent, eternal being is fueled by several sources. The most obvious is the appeal to natural and historical revelation. Conscience (immediate natural revelation) and biblical revelation both point to the personality of God.

Even apart from appeals to such revelation there have also been appeals to both cosmological and theological reasoning to ascribe personality to God. The cosmological argues from the presence of personality in this world to the ultimate cause of personality. René Descartes, for example, argues that there cannot be more in the effect than there is in the cause. There can be less, but not more. This leads to another question: Can there be an impersonal cause of personality ultimately? The cosmological builds its case on the nature of personality. If, as Edmund Husserl argued, the definitive characteristic of personality is intentionality, then the complex argument deepens.

To act with intention is to act with knowledge and will. There can be no intention without mind operating somewhere. Flowers may respond to external stimuli, but that is not the same thing as an *intentional* act. An intentional act is conscious and volitional. If it is not intentional, then by the nature of the case it must be unintentional. If it is unintentional, then it must be considered an accident.

The question then becomes, Can intentionality come to pass unintentionally? Accidental intention is a contradiction in terms. If true intentionality exists among contingent beings, we are forced to ask, Does this intentionality come about unintentionally?

Because of questions like these both Immanuel Kant and David Hume were more impressed or bothered by the teleological argument from design. Both agreed

From painting by Frans Hals, in the gallery of the Louvre

René Descartes
(1596–1650)

that the world, as we perceive it, displays uncanny signs of order, organization, and purpose. They agreed it testifies to the work of an architect. The world, as Carl Sagan notes, is "cosmos," not "chaos."

Some have sought to escape this by arguing that the "order" of the "cosmos" is not objective, but subjective. That is, there is no real order "out there." The order is in our minds, and we impose it on an unordered world by our scientific penchant for classification and organization.

Even if the subjective theory of imposed order were true, however, one must still account for the presence in our minds of real order. Order would still exist in the universe even if it were only in our minds. We would still have the presence of intentionality.

The question of God's personality goes beyond the scope of this work. The critical issue remains: reason demands that there be self-existing, eternal being.

The demand of reason rests on two assumptions, both of which are integral to rationality: the law of noncontradiction and the law of causality. We have already seen that the law of causality is an extension of the law of noncontradiction, but it is a significant extension.

Since the cosmological argument for self-existent eternal being rests so heavily on the law of causality, it is no surprise that attacks against such being often focus on causality. It was at the heart of both Kant's and Hume's critiques of the classical arguments for God's existence. We remember Russell's observation that "the philosophers and men of science have got going on cause, and it has not anything like the vitality it used to have."

We must therefore take a closer look at causality, particularly with respect to Hume and his predecessors.

No Chance
in the World

"Chance, when strictly examined, is a mere negative word, and means not any real power which has anywhere a being in nature."

•

David Hume

lassical philosophers distinguished among various types of causes. For example, Aristotle mentioned material cause, formal cause, final cause, instrumental cause, efficient cause, and sufficient cause. (See fig. 10.)

A *material cause* may be defined as that *out of which* something is made. In the case of a statue the material cause would be the stone out of which the figure is shaped.

The *formal cause* is the design, pattern, blueprint, or idea that is followed in the process. It determines the *form* the object will take. In the case of a statue it may be a sketch that is made before the sculpting itself begins.

The *final cause* is the *purpose* for which a thing is made. A statue may be commissioned, for example, to adorn a church or to beautify a garden. The ultimate end is in view with respect to the final cause. The final cause is expressed in terms of teleology.

The *instrumental cause* refers to the means or instrument by which a thing is made. In the case of a statue the instrumental cause is the *chisel* the sculptor uses to shape the stone.

The *efficient cause* is the main or chief causal agent that brings about the effect. The efficient cause is *necessary* for the thing to be made. In the case of a statue it is the sculptor himself.

The *sufficient cause* is a cause that is equal to the task. It is *able* to bring about the desired effect.

A further distinction in causality arose with the debate, particularly in the seventeenth century, concerning the relationship between the first and ultimate cause and the later, proximate causes, or the relationship between the causal power of God and the causal agency of contingent beings. This distinction was marked with the terms *primary causality* and *secondary causality*. Secondary causality, though real and significant, depends for its power ultimately on the primary cause. Since created being is dependent, contingent, and derived, it always depends on self-existent being for its own existence and power.

The question of the causal relationship between contingent things as well as between God and contingent things was a burning issue in the seventeenth-century Age of Reason. Several different approaches were taken to this question before David Hume applied his critique of causality.

Descartes and Malebranche

René Descartes, recognized as the father of seventeenth-century rationalism, sought vigorously for clear and distinct ideas whose truth was indubitable. Fol-

Fig. 10
Aristotle's Six Causes

	Definition	Example
Material cause	That out of which something is made.	The stone out of which a statue is carved.
Formal cause	The design or idea followed in the process of making something.	A sketch made by the sculptor as a pattern for the sculpture.
Final cause	The purpose for which something is made.	The reason why the sculptor is doing the sculpture.
Instrumental cause	The means or instrument by which something is made.	The sculptor's chisel.
Efficient cause	The chief agent causing something to be made.	The sculptor.
Sufficient cause	A cause equal to the task of causing the thing to be made.	A person capable of sculpting.

lowing a rigorous process of systematic doubting, he arrived at his first principle of his own existence: *Cogito ergo sum.* In this famous formula Descartes arrived at self-consciousness as a foundation for further deduction. Descartes reached a point of intense struggle when he analyzed the question of the nature of created reality. He focused on the question of the nature of mind and matter and on their relationship. He saw the chief attribute or characteristic of mind as *thought.* The chief attribute of matter is *extension.*[1]

It is one thing to ponder how one thought can give rise to (or cause) another thought. It is quite another to consider how a thought, which is nonextended, can give rise to (or cause) a physical action, which is extended. The question became, How does extended reality interact with nonextended reality? How do body and mind interact?

I have a sense that I can think about picking up a pen and then proceed to use my hand to pick up the pen. Is this process entirely physical, including thought itself, or is there some mysterious process of interaction between mind and body?

Descartes' disciples Nicolas de Malebranche and Arnold Geulincx modified Descartes' theories of interactionism into a theory known as *occasionalism.*

Occasionalism represented an attempt to resolve the question of primary and secondary causality. It was also an effort to preserve the necessary involvement of God's providence within the order of created reality. The idea that the created world is a self-contained, closed, mechanistic unity, functioning strictly via "natural causes" of physical interaction between bodies, was anathema to these men.

In Malebranche the pole of secondary causality is swallowed up by the primary causality of God. James Collins says that "Malebranche's occasionalism consisted in a denial of finite causality and an ascription of all real causality to God alone within a framework of finite 'occasions.'"[2] For Malebranche and the occasionalists the relationship between mind and body, thought and action, is more a case of *parallelism* than of *interaction.*

Frederick Copleston summarizes Malebranche:

God, therefore, is the one and only true cause.... Certainly, there is a natural order in the sense that God has willed, for example, that *A* should always be followed by *B*, and this order is constantly preserved because God has willed that it should be preserved. To all outward appearance, therefore, it seems that *A* causes *B*. But metaphysical reflection shows that *A* is simply an occasional cause. The fact that on the occurrence of event *A* God always causes event *B* does not show that *A* is a true cause of *B*. ... this does not mean that causality in general is nothing more than regular sequence. It means that natural causes are not true causes and that the only true cause is a supernatural agent, God.[3]

The significance of occasionalism for Hume's later critique of causality is great. Malebranche spoke of the "outward appearances" of causality, that is, what *is perceived* as a causal connection but which is not *the* causal connection. Yet he was careful not to reduce causality to mere sequence. The occasionalists did not deny the reality of causality. Their concern was in the *locus* and identification of the true causal power.

Spinoza and Leibniz

Other attempts to define the interaction or apparent causal relationships between entities were proffered by Gottfried Leibniz and Baruch Spinoza.

Spinoza gave a different spin to the concept of causality, which he rooted and grounded in his concept of *substance.* Roger Scruton summarizes the distinctive character of Spinoza's substance philosophy:

> A substance must be intelligible apart from all relations with other things. Hence . . . a substance cannot enter into relations and in particular cannot enter into causal relations. To the extent that it does so, it has to be explained in terms of, and hence "conceived through," other things. A substance cannot therefore be produced by anything else, it must therefore be its own cause (*causa sui*). By definition nothing can be such unless its essence involves existence.[4]

For Spinoza, substance is all. Substance contains attributes and modes. An attribute may be defined as a particular manifestation of substance. A mode is a particular manifestation of an attribute. The bottom line is that all is substance and substance is all, differing only in attributes and modes. For Spinoza all cause is God because everything is God. God did not create the world; God is the world.

Wilhelm Windelband remarks:

> Spinoza always expresses his conception of real dependence, of causality, by the word "follow" (*sequi, consequi*) and by the addition, "as from the definition of a triangle the equality of the sum of its angles to two right angles follows." The dependence of the world upon God is, therefore, thought as a *mathematical consequence.* This conception of the causal relation has thus completely stripped off the empirical mark of "producing" or "creating" which played so important

a part with the Occasionalists, and replaces the perceptional idea of active operation with the *logico-mathematical relation of ground and consequent [or reason and consequent; Grund und Folge]*. Spinozism is a consistent identification of the relation of cause and effect with that of ground and consequent.[5]

Spinoza writes:

... the order of concatenation of things is a single order, whether Nature is conceived under one or the other attribute; it follows therefore that the order of the action and passions of our body is simultaneous in nature with the order of the actions and passions of the mind. . . . Now all these things clearly show that the decision of the mind, together with the appetite and determination of the body, are simultaneous in nature, or rather that they are one and the same thing, which, when it is considered under the attribute of thought and explained in terms of it, we call decision, and when considered under the attribute of extension, and deduced from the laws of motion and rest, we call causation.[6]

Gottfried Leibniz also sought a metaphysical explanation for the phenomenon of causality. His intricate system of monadology gave rise to his famous theory of the *law of preestablished harmony*. Leibniz's "monads" cannot interact with each other. They merely "appear" to interact in the phenomenal world. What we examine in the empirical activity of science is the world as it appears to us. Beneath the surface of appearance is a metaphysical reality that cannot be perceived. Causal interaction is a result of a harmony of monads that is established in eternity by God. Windelband remarks:

> The pre-established harmony—this relationship of substances in their Being and life—needs, however, a *unity* as the ground of its explanations, and this can be sought only in the central monad. God, who created the finite substances, gave to each its own content in a particular grade of representative intensity, and thereby so arranged all the monads that they should harmonise throughout. And in this necessary process in which their life unfolds, they realise the end of the creative Universal Spirit in the whole mechanical determination of the series of their representations.[7]

Hume's Critique

We see, then, that prior to Hume's watershed critique of causality, a huge dynamic of philosophical inquiry into the question of causality had been played out. From Descartes' *interactionism* to *occasionalism* to Spinoza's *substance philosophy* to Leibniz's *law of preestablished harmony,* we see disparate views of causality contending with and challenging one another. It takes Hume to lay his ax at the root of the tree.

David Hume, in his skeptical analysis of epistemology and metaphysics, is widely considered as having buried science in the "graveyard of empiricism." His critique was the occasion for Immanuel Kant's awakening from his "dogmatic slumbers" and attempting to rescue science from skepticism. Kant remained agnostic with respect to metaphysical inquiry, assigning concepts such as God, the self, and essences to the unknowable realm of *noumena,* but he sought to rescue a place for empirical science at the level of the phenomenal realm.

David Hume
(1711–1776)

Hume's analysis of causality was central to this watershed event in theoretical thought. When Bertrand Russell remarked that "cause is not quite what it used to be" because "the philosophers and the men of science have got going on cause," he probably had Hume's critique of causality in mind.

In *An Enquiry Concerning Human Understanding* Hume distinguished between two kinds of objects: *relations of ideas* studied in geometry, algebra, and arithmetic; and *matters of fact.* This followed closely the classical distinction between the examination of the formal and that of the material. Hume remarks: "All reasonings concerning matter of fact seem to be founded on the relation of *Cause and Effect.* By means of that relation alone we can go beyond the evidence of our memory and senses."[8]

In the analysis of matters of fact Hume was determined to push the question of the nature and credulity of the evidence for them. He asserts:

> I shall venture to affirm, as a general proposition, which admits of no exception, that the knowledge of this relation is not, in any instance, attained by reasonings *a priori*; but arises entirely from experience. . . .[9]

He goes on to say:

> No object ever discovers, by the qualities which appear to the senses, either the causes which produced it, or the effects which will arise from it; nor can our rea-

son, unassisted by experience, ever draw any inference concerning real existence and matter of fact. . . .
*causes and effects are discoverable, not by reason but
by experience. . . .*[10]

So far, at least, Hume has not denied the law of
causality. He is not speaking of uncaused effects or
effectless causes. He is speaking instead of knowledge
or discovery of the facts regarding causes and effects
in the empirical world. He still speaks of "causes which
produce" and effects that "arise from" such causes.

Early in this essay he introduces his famous illustration of the action of billiard balls. When we observe
a game of pool or billiards, we observe the following
actions: (1) a player strikes the cue ball with the tip of
the cue stick; (2) the cue ball, which was formerly at
rest, begins to move across the table; (3) the cue ball
then strikes the object ball, and the object ball rolls
into the corner pocket.

In this scenario we observe change. We observe
motion. We assume that the changes from rest to
motion and back to rest are caused by some force. We
assume that the force is initiated by the pool player
who moves the stick, which moves the cue ball, which
moves the object ball. We *assume* a causal chain or
nexus in this experience.

Hume ponders the assumptions we bring to the pool
table:

We fancy, that were we brought on a sudden into this
world, we could at first have inferred that one Billiard-

ball would communicate motion to another upon impulse; and that we needed not to have waited for the event, in order to pronounce with certainty concerning it. . . .

. . . The mind can never possibly find the effect in the supposed cause, by the most accurate scrutiny and examination. For the effect is totally different from the cause, and consequently can never be discovered in it. Motion in the second Billiard-ball is a quite distinct event from motion in the first; nor is there anything in the one to suggest the smallest hint of the other.[11]

Hume insists that since every effect is a distinct event from its cause, it cannot be discovered in the cause. (Again notice that we're still speaking of causes and effects, but we're limiting discovery of the precise relationship between them.)

Cause and Custom

Hume moves on. He then argues that even after we have *experience* (a posteriori) of cause and effect, "our conclusions from that experience are *not* founded on reasoning, or any process of the understanding."[12] He summarizes by saying:

We have said that all arguments concerning existence are founded on the relation of cause and effect; that our knowledge of that relation is derived entirely from experience; and that all our experimental conclusions proceed upon the supposition that the future will be conformable to the past. To endeavour, therefore, the proof of this last supposition by probable arguments, or arguments regarding existence, must be evidently going in a circle, and taking that for granted, which is the very point in question.[13]

If we were to observe the rain falling ten thousand times and notice that each time it falls the grass gets wet, we would be inclined to suppose that there is a necessary connection, indeed a causal connection, between the rain's falling and the grass's becoming wet. We conclude that falling rain causes wet grass.

Since we do not have a direct and immediate perception of a causal connection between rain and wet grass, all we can do with respect to predicting the future is to speak in terms of probabilities. If it rains tomorrow, the probability quotient that the grass will get wet is enormous. But it is not certain that the one will be followed by the other.

The basis for predicting the future is not rational analysis but thinking based on custom. Hume explains:

> Suppose a person, though endowed with the strongest faculties of reason and reflection, to be brought on a sudden into this world; he would, indeed, immediately observe a continual succession of objects, and one event following another; but he would not be able to discover anything farther. He would not, at first, by any reasoning, be able to reach the idea of cause and effect; since the particular powers, by which all natural operations are performed, never appear to the senses; nor is it reasonable to conclude, merely because one event, in one instance, precedes another, that therefore the

one is the cause, the other the effect. Their conjunction may be arbitrary and casual.[14]

Hume is simply applying the informal logical fallacy *Post hoc ergo propter hoc*. Simply because one thing, or one event, follows another in temporal procession, this does not mean that one is caused by the other. It is like the rooster and the sunrise. The rooster crows and then the sun rises, but this does not mean that the rooster's crowing has caused the sun to rise.

Hume asserts that what we perceive in such sequences is not necessarily a causal relation but a *customary relationship,* or a relationship of *contiguity.* Hume says this regarding custom:

> Custom, then, is the great guide of human life. It is that principle alone which renders our experience useful to us, and makes us expect, for the future, a similar train of events with those which have appeared in the past. Without the influence of custom, we should be entirely ignorant of every matter of fact beyond what is immediately present to the memory and senses.[15]

It is important to note in passing that all the while Hume is distinguishing between a causal connection and a customary relationship, he remains constrained to explain why we have the assumption of causal connection. He is searching for a "reason" or even "cause" for the assumption of causal connection. He is still thinking causally. With respect to custom he says that it "renders," "makes," and "influences." That is, custom yields something that in ordinary language is called an "effect." Indeed later Hume uses this language:

> Custom is that principle, by which this correspondence has been effected. . . .[16]

Chance and Causation

Hume returns to the question of probability and makes use of the term *chance*. He is careful, however, not to confuse chance with any ontological reality:

> Though there be no such thing as *Chance* in the world; our ignorance of the real cause of any event has the same influence on the understanding, and begets a like species of belief or opinion.[17]

Hume clearly acknowledges that chance has no existence: there is "no such *thing* as *chance* in the world." He also speaks of *the real cause* of any event. We may be ignorant of the real cause, but it emphatically does not follow that there is no real cause.

We know that chance can never be the real cause of anything because there is no such *thing (res)* in the world as chance. Speaking of "chance" in a causal way is begotten from ignorance of real causes. Chance is an *unreal* cause, which is no cause.

In his next paragraph Hume continues to speak of chance:

There is certainly a probability, which arises from a superiority of chances on any side; and according as this superiority increases, and surpasses the opposite chances, the probability receives a proportionable increase, and begets still a higher degree of belief or assent to that side, in which we discover the superiority.[18]

Here Hume speaks of chance in the mathematical sense of probability, but not in any ontological or causal sense.

At this point in *Enquiry* Hume returns to the question of necessary connection:

When we look about us towards external objects, and consider the operation of causes, we are never able, in a single instance, to discover any power or necessary connexion; any quality, which binds the effect to the cause, and renders the one an infallible consequence of the other. We only find, that the one does actually, in fact, follow the other. The impulse of one billiard-ball is attended with motion in the second. This is the whole that appears to the *outward* senses.[19]

After discussing the theories of the occasionalists, Hume concludes:

We are ignorant, it is true, of the manner in which bodies operate on each other: their force or energy is entirely incomprehensible: but are we not equally ignorant of the manner or force by which a mind, even the supreme mind, operates either on itself or on body?[20]

This agnosticism with respect to our ability to perceive particular causes or of necessary connection has implications for science. Hume avers:

The only immediate utility of all sciences, is to teach us, how to control and regulate future events by their causes.[21]

Hume's Critique and Science

Since in Hume's view we are ignorant of necessary connections, what does this mean for science? If the relationship between events is customary but not causal, are we still able to predict or manipulate our environment? Surely Hume would grant that we can still make such predictions along the lines of probabilities, even though we remain uncertain about particular causes.

For practical purposes, if my grass starts to turn brown, I can still water it and hope that it will turn green again. The ultimate cause may rest in some preestablished harmony, metaphysical substance, or occasionalist intervention by God. One thing I will not do is leave the future of my grass up to chance. I can at least hope that the application of water will yield a good result. I have no hope that chance will do the job.

As we have already seen, David Hume did not fall back on chance as some mysterious force that explains the behavior of the empirical world. Again Hume strongly asserts:

> It is universally allowed that nothing exists without a cause of its existence, and that chance, when strictly examined, is a mere negative word, and means not any real power which has anywhere a being in nature.[22]

It is the "strict" examination of chance that is the subject of this book. Hume is correct when he says that chance has no power because it has no being in nature.

Hume states a truism when he says that "nothing exists without a cause of its existence." His analysis of causality does not annihilate or disturb the law of causality. He focuses on the limits of empirical perception and on our inability to arrive at demonstrative knowledge of necessary connection. This makes no dent in the formal law of causality, which is analytically true. It is true by definition. It carries no synthetic baggage. The law of causality has no concrete content. It is theoretically possible that there are no real causes and/or no real effects in the universe. Though I believe there are both causes and effects, my belief does not make it so.

What is unassailably true is that if there are real effects there must be real causes, and if there are real causes there must be real effects.

Science and philosophy use the formal law of causality in their spheres of investigation as an extension of the law of noncontradiction. Assumptions are made about effects, and then searches are undertaken for sufficient causes of those effects. Physicians search for "causes" of diseases and causes of medical healing.

When something is assumed to be an effect, it is usually done so by virtue of a perceived change in its state. Change, as Aristotle observed, may be understood as a type of motion. Generation and decay, aging, movement, and the like are all forms of change and/or motion. They are the ingredients of what we call contingency. When something changes its state, we naturally ask, "What changed it?" When something moves, we ask, "What moved it?"

The law of inertia is a case in point. It is usually defined this way: A body at rest tends to remain at rest unless acted on by an outside force. A body in motion tends to remain in motion unless acted on by an outside force.

I had a vivid existential experience of the law of inertia in the 1993 Alabama Amtrak train wreck. Asleep in the last car of the train, I was suddenly awakened by the noise and force of the wreck. I awakened in midair as I was hurled from my berth across the room and into the wall. My rest was disturbed by an outside force. When the train stopped, I remained in motion until my motion was arrested by the wall. I don't know what was the real cause of my awakening from dogmatic slumber. It may have been the result of preestablished harmony.

One thing I know for sure. The train did not wreck by chance. It may have been an accident, but it had a real cause. As I am writing these lines, there is a news report on television documenting the official hearings

on the accident, invoking the testimony and culpability of the barge company whose runaway barges
slammed into the bridge, after which the rails were
moved. In seeking to locate the responsibility for this
accident, none of the parties involved is appealing to
the causal power of chance as the real culprit.

In a significant way Hume's study of causality antic-
ipated Werner Heisenberg's theory of indeterminacy.
The mysterious behavior of quantum particles posits
at least a temporary limit to our powers of perceiving
the real cause(s) of motion or change. But they do not
give us license to adopt an irrational, unscientific view
of chance as a causal agent, force, or power.

Chance as a real force is a myth. It has no basis in
reality and no place in scientific inquiry. For science
and philosophy to continue the advance in knowledge,
chance must be demythologized once and for all.

Notes

Preface

1. Immanuel Velikovsky, *Worlds in Collision* (Garden City, NY: Doubleday, 1950).

2. James Gleick, *Chaos: Making a New Science* (New York: Viking, 1987).

Chapter 1, "The Soft Pillow"

1. Stanley L. Jaki, *God and the Cosmologists* (Washington, DC: Regnery Gateway, 1989), p. 167. From Arthur Koestler, *Darkness at Noon*, trans. Daphne Hardy (New York: Bantam, 1941), p. 149. Hardy translates the dependent clause as follows: "As long as chaos dominates the world . . ."

2. Jaki, *God and the Cosmologists*, p. 149. From Pierre Delbet, *La science et la réalité* (Paris: Flammarion, 1913), p. 238.

3. Jaki, *God and the Cosmologists*, p. 150.

4. John H. Gerstner, *The Rational Biblical Theology of Jonathan Edwards*, vol. 1 (Powhatan, VA: Berea Publications, 1991), p. 123. From "Of Being," ms. 312, p. 9.

5. Jaki, *God and the Cosmologists*, p. 150.

6. Ibid., p. 161. From George Wald, "The Origin of Life," in *The Physics and Chemistry of Life,* A *Scientific American* Book (New York: Simon and Schuster, 1955), p. 12.

7. Jaki, *God and the Cosmologists*, pp. 163, 164.

Chapter 2, "The Mask of Ignorance"

1. Stanley L. Jaki, *God and the Cosmologists* (Washington, DC: Regnery Gateway, 1989), p. 142. From Paul Janet, *Final Causes,* 2d ed., trans. William Affleck (New York: Scribner's, 1891), p. 19.

2. Ibid., p. 143.

3. Ibid., p. 145.

4. Ibid. From Jacques Bossuet, *Discours sur l'histoire universelle,* 3.8. For another translation see *Discourse on Universal History,* trans. Elborg Forster (Chicago: University of Chicago, 1976), p. 374.

5. Jaki, *God and the Cosmologists*, p. 145. David Hume's precise words follow: "Though there be no such thing as *Chance* in the world; our ignorance of the real cause of any event has the same influence on the understanding, and begets a like species of belief or opinion." *An Enquiry Concerning Human Understanding* (Buffalo: Prometheus, 1988), p. 55 (section 6). Italics his.

6. Jaki, *God and the Cosmologists*, pp. 145–46. From Voltaire, *Dictionnaire philosophique,* in *Oeuvres complètes de Voltaire* (Paris: Garnier Frères, 1877–85), 17:478.

7. Jaki, *God and the Cosmologists*, p. 146. From Claude Helvétius, *De l'homme* 1.8, in *Oeuvres complètes* (Paris: Lepetit, 1818), 2:33.

8. Jaki, *God and the Cosmologists*, p. 146. From T. H. Huxley's reminiscences on the publication of *Origin of Species,* in *The Life and Letters of Charles Darwin,* ed. Francis Darwin, 2 vols. (New York: Appleton, 1896), 1:553.

9. Jaki, *God and the Cosmologists*, pp. 146–47. From Charles Darwin, letter to J. D. Hooker (1870), in *More Letters of Charles Darwin: A Record of His Work in a Series of Hitherto Unpublished Letters,* ed. Francis Darwin and A. C. Seward, 2 vols. (New York: Appleton, 1903), 1:321.

10. Jaki, *God and the Cosmologists*, p. 148. From Charles S. Peirce, "The Order of Nature," *Popular Science Monthly* 13 (June 1878): 207.

Chapter 3, "A Quantum Leap"

1. Timothy Ferris, *Coming of Age in the Milky Way* (New York: Morrow, 1988), p. 291.

2. Ibid.

3. Ibid., p. 30.
4. Ibid., pp. 30–31.
5. Ibid., pp. 286, 288. Italics his.
6. Ibid., p. 288. Italics his.
7. R. C. Sproul, John H. Gerstner, and Arthur W. Lindsley, *Classical Apologetics: A Rational Defense of the Christian Faith and a Critique of Presuppositional Apologetics* (Grand Rapids: Zondervan, 1984).
8. William Poundstone, *Labyrinths of Reason: Paradox, Puzzles, and the Frailty of Knowledge* (New York: Doubleday, 1988), p. 41.
9. John Boslough, *Stephen Hawking's Universe* (New York: Morrow, 1985), p. 57; (New York: Avon, 1989), p. 48. Fritjof Capra, *The Tao of Physics: An Exploration of the Parallels between Modern Physics and Eastern Mysticism,* 3d ed. (Boston: Shambhala, 1991); Michael Talbot, *Mysticism and the New Physics* (New York: Bantam, 1981).

Chapter 4, "The Voice of Reason"

1. Carl Sagan, *Cosmos* (New York: Random, 1980), p. 347; (New York: Ballantine, 1985), p. 289.
2. Ibid.
3. Nigel Calder, *Einstein's Universe* (New York: Viking, 1979), p. 140; (New York: Penguin, 1980), p. 234.
4. Timothy Ferris, *Coming of Age in the Milky Way* (New York: Morrow, 1988), p. 290.
5. Ibid. Albert Einstein to Max and Hedwig Born, 29 April 1924, *The Born-Einstein Letters: Correspondence between Albert Einstein and Max and Hedwig Born, from 1916 to 1955, with Commentaries by Max Born,* trans. Irene Born (New York: Walker, 1971), p. 82. Italics Einstein's.
6. Ferris, *Coming of Age in the Milky Way,* p. 290. Albert Einstein to Max Born, 4 December 1926, *The Born-Einstein Letters,* p. 91 (italics Einstein's); Einstein to Born, 3 March 1947, *The Born-Einstein Letters,* p. 158.
7. Calder, *Einstein's Universe* (Viking), p. 141; (Penguin), p. 235.
8. Ibid.
9. Ibid., (Viking), p. 141; (Penguin), p. 236.
10. Stanley L. Jaki, *God and the Cosmologists* (Washington, DC: Regnery Gateway, 1989), pp. 164–65.
11. Ibid., p. 165.
12. Ibid., pp. 165–66. The four quotations are from James Gleick, *Chaos: Making a New Science* (New York: Viking, 1987), pp. 5, 269, 314, 197. The fourth quoted phrase refers to Theodor Schwenk, *Sensitive Chaos* (New York: Schocken, 1976).
13. William Manchester, *A World Lit Only by Fire: The Medieval Mind and the Renaissance: Portrait of an Age* (Boston: Little, Brown, 1992), p. 290.

14. Ibid.
15. Ibid.
16. Ibid., p. 291.
17. Ferris, *Coming of Age in the Milky Way*, pp. 288–89.
18. Roger Penrose, *The Emperor's New Mind: Concerning Computers, Minds, and the Laws of Physics* (New York: Oxford University, 1989), p. 230.
19. Ibid., p. 255. Italics his.
20. Ibid., p. 256.

Chapter 5, "Light and *the* Light"

1. Stanley L. Jaki, *God and the Cosmologists* (Washington, DC: Regnery Gateway, 1989), p. 152. From Abraham Pais, "Einstein and the Quantum Theory," *Reviews of Modern Physics* 51 (1979): 907.
2. Jaki, *God and the Cosmologists*, p. 118.
3. Ibid.
4. Ibid.

Chapter 6, "Framing the Question"

1. Timothy Ferris, *The Mind's Sky: Human Intelligence in a Cosmic Context* (New York: Bantam, 1992), pp. xii–xiii.
2. Ibid., p. 3.
3. Ibid., p. 4. Italics his.
4. Ibid., p. 5. Italics his.
5. Ibid., p. 7. Italics mine.
6. Allan Bloom, *The Closing of the American Mind* (New York: Simon and Schuster, 1987).
7. Ferris, *The Mind's Sky*, p. 48.
8. Ibid., pp. 49–50. Italics his.

Chapter 7, "The Policeman of Science"

1. Timothy Ferris, *Coming of Age in the Milky Way* (New York: Morrow, 1988), pp. 65–66. See Nicolaus Copernicus, *On the Revolutions of the Heavenly Spheres*, trans. Charles Glenn Wallis, book 1, section 1. In Robert Maynard Hutchins, ed., *Great Books of the Western World*, vol. 16, *Ptolemy, Copernicus, Kepler* (Chicago: Encyclopaedia Britannica, 1952), p. 511.

2. Ferris, *Coming of Age in the Milky Way*, pp. 78–79. From Johannes Kepler, *Astronomia Nova*, XL. Alexandre Koyré, *The Astronomical Revolution*, trans. R. E. W. Maddison (Ithaca, NY: Cornell University, 1973), p. 231.

3. Ferris, *Coming of Age in the Milky Way*, p. 79. Italics his.

4. Ibid., p. 77.

5. Ronald H. Nash, *Worldviews in Conflict: Choosing Christianity in a World of Ideas* (Grand Rapids: Zondervan, 1992), p. 82.

Chapter 8, "Cosmos or Chaos?"

1. J. P. Moreland, *Christianity and the Nature of Science* (Grand Rapids: Baker, 1989), p. 122.

2. Ibid., p. 124.

3. William Poundstone, *Labyrinths of Reason: Paradox, Puzzles, and the Frailty of Knowledge* (New York: Doubleday, 1988), p. 16.

4. Carl Sagan, *Cosmos* (New York: Random, 1980), p. 29; (New York: Ballantine, 1985), p. 18.

Chapter 9, "A Being without a Cause"

1. Bertrand Russell, *"Why I Am Not a Christian" and Other Essays on Religion and Related Subjects,* ed. Paul Edwards (New York: Simon and Schuster, 1957), pp. 3–4.

2. Ibid., p. 7.

3. R. C. Sproul, *The Psychology of Atheism* (Minneapolis: Bethany, 1974). Reprinted as *If There Is a God, Why Are There Atheists?* (Minneapolis: Bethany, 1978; Wheaton: Tyndale, 1988).

Chapter 10, "No Chance in the World"

1. René Descartes, *The Meditations Concerning First Philosophy*, "Fifth Meditation," in *"Discourse on Method" and "Meditations,"* ed. and trans. Laurence J. Lafleur, Library of Liberal Arts (Indianapolis: Bobbs-Merrill, 1960), p. 118.

2. James Collins, *God in Modern Philosophy* (Chicago: Regnery, 1959), p. 86.

3. Frederick Copleston, *A History of Philosophy*, vol. 4, *Descartes to Leibniz* (Garden City, NY: Image, 1963), pp. 196–97. See Nicholas de Malebranche, *Entretiens sur la métaphysique*, 7.9–10.

4. Roger Scruton, *From Descartes to Wittgenstein: A Short History of Modern Philosophy* (Boston: Routledge and Kegan Paul, 1981), p. 53.

5. Wilhelm Windelband, *A History of Philosophy*, trans. James H. Tufts, 2d ed., 2 vols. (New York: Harper and Row, 1958), 2:418. Italics his.

6. Scruton, *From Descartes to Wittgenstein*, p. 60.

7. Windelband, *A History of Philosophy*, pp. 424–25. Italics his.

8. David Hume, *An Enquiry Concerning Human Understanding*, Great Books in Philosophy (Buffalo: Prometheus, 1988), p. 29 (section 4, part 1). Italics his.

9. Ibid., p. 30 (sec. 4, pt. 1).

10. Ibid. Italics his.

11. Ibid., p. 31 (sec. 4, pt. 1).

12. Ibid., p. 34 (sec. 4, pt. 2). Italics his.

13. Ibid., p. 37 (sec. 4, pt. 2).

14. Ibid., pp. 42–43 (sec. 5, pt. 1).

15. Ibid., p. 45 (sec. 5, pt. 1).

16. Ibid., p. 53 (sec. 5, pt. 2).

17. Ibid., p. 55 (sec. 6). Italics his.

18. Ibid.

19. Ibid., p. 60 (sec. 7, pt. 1). Italics his.

20. Ibid., p. 68 (sec. 7, pt. 1).

21. Ibid., p. 72 (sec. 7, pt. 2).

22. Ibid., p. 88 (sec. 8, pt. 1).

Bibliography

Aristotle. *Aristotle: Selections.* Edited by W. D. Ross. Modern Student's Library: Philosophy Series. New York: Scribner's, 1927.

Asimov, Isaac. *Understanding Physics.* 3 vols. New York: Walker, 1966.

Boslough, John. *Stephen Hawking's Universe.* New York: Morrow, 1985.

Calder, Nigel. *Einstein's Universe.* New York: Viking, 1979.

Casserley, J. V. Langmead. *The Christian in Philosophy.* New York: Scribner's, 1951.

Clark, Gordon H. *The Philosophy of Science and Belief in God.* Nutley, NJ: Craig, 1964.

Collins, James. *God in Modern Philosophy.* Chicago: Regnery, 1959.

Copleston, Frederick. *A History of Philosophy.* Vol. 1, *Greece and Rome.* Garden City, NY: Image, 1962.

_____. *A History of Philosophy.* Vol. 4, *Descartes to Leibniz.* Garden City, NY: Image, 1963.

_____. *A History of Philosophy.* Vol. 5, *Modern Philosophy: The British Philosophers.* Garden City, NY: Image, 1964.

Craig, William Lane. *The Cosmological Argument from Plato to Leibniz.* New York: Barnes and Noble, 1980.

Descartes, René. *"Discourse on Method" and "Meditations."* Edited and translated by Laurence J. Lafleur. Library of Liberal Arts. Indianapolis: Bobbs-Merrill, 1960.

Dulles, Avery R., James M. Demske, and Robert J. O'Connell. *Introductory Metaphysics: A Course Combining Matter Treated in Ontology, Cosmology, and Natural Theology.* New York: Sheed and Ward, 1955.

Ferré, Frederick. *Language, Logic, and God.* New York: Harper, 1961.

Ferris, Timothy. *Coming of Age in the Milky Way.* New York: Morrow, 1988.

_____. *The Mind's Sky: Human Intelligence in a Cosmic Context.* New York: Bantam, 1992.

Gerstner, John H. *The Rational Biblical Theology of Jonathan Edwards.* Vol. 1. Powhatan, VA: Berea, 1991.

Gilson, Étienne. *History of Christian Philosophy in the Middle Ages.* New York: Random, 1955.

_____. *Reason and Revelation in the Middle Ages.* New York: Scribner's, 1938.

Gleick, James. *Chaos: Making a New Science.* New York: Viking, 1987.

Hume, David. *An Enquiry Concerning Human Understanding.* Great Books in Philosophy. Buffalo: Prometheus, 1988.

_____. *Dialogues Concerning Natural Religion.* Edited by Henry D. Aiken. Hafner Library of Classics. New York: Hafner, 1957.

Jaki, Stanley L. *God and the Cosmologists.* Washington, DC: Regnery Gateway, 1989.

Jones, W. T. *A History of Western Philosophy.* Vol. 3, *Hobbes to Hume.* 2d ed. New York: Harcourt Brace Jovanovich, 1969.

Kant, Immanuel. *Critique of Pure Reason.* Translated by F. Max Muller. 2d ed. New York: Doubleday, 1961.

Kaufmann, Walter Arnold, ed. *Philosophic Classics: Basic Texts.* Vol. 2, *Bacon to Kant.* Prentice-Hall Philosophy Series. Englewood Cliffs, NJ: Prentice-Hall, 1961.

Luijpen, W. A. M. *Fenomenologie en Atheisme.* 2d ed. Utrecht: Spectrum, 1967.

Manchester, William. *A World Lit Only by Fire: The Medieval Mind and the Renaissance: Portrait of an Age.* Boston: Little, Brown, 1992.

Moreland, J. P. *Christianity and the Nature of Science*. Grand Rapids: Baker, 1989.

Mourelatos, Alexander P. D., ed. *The Pre-Socratics: A Collection of Critical Essays*. Modern Studies in Philosophy. Garden City, NY: Anchor, 1974.

Nash, Ronald H. *Worldviews in Conflict: Choosing Christianity in a World of Ideas*. Grand Rapids: Zondervan, 1992.

Penrose, Roger. *The Emperor's New Mind: Concerning Computers, Minds, and the Laws of Physics*. New York: Oxford University, 1989.

Poundstone, William. *Labyrinths of Reason: Paradox, Puzzles, and the Frailty of Knowledge*. New York: Doubleday, 1988.

Ross, Hugh. *The Creator and the Cosmos: How the Greatest Scientific Discoveries of the Century Reveal God*. Colorado Springs: NavPress, 1993.

Russell, Bertrand. *"Why I Am Not a Christian" and Other Essays on Religion and Related Subjects*. Edited by Paul Edwards. New York: Simon and Schuster, 1957.

Sagan, Carl. *Cosmos*. New York: Random, 1980.

Scruton, Roger. *From Descartes to Wittgenstein: A Short History of Modern Philosophy*. New York: Harper and Row, 1981.

Windelband, Wilhelm. *A History of Philosophy*. Translated by James H. Tufts. 2d ed. 2 vols. New York: Harper and Row, 1958.

Index of Scholars

Adler, Mortimer, xv, 62

Aquinas. *See* Thomas Aquinas

Aristotle *(Greek philosopher, 384–322 B.C.)*, 16, 24, 30, 38, 66, 71, 109, 111, 114, 116, 128, 136, 138, 195, 213

Berkeley, George (Bishop) *(Irish philosopher, 1685–1753)*, 94, 102, 110

Bloom, Allan *(American philosopher, 1930–1992)*, 113

Bohr, Niels *(Danish physicist, 1885–1962)*, 16–17, 54, 57–59, 60–61, 62, 73, 97, 144–45

Born, Max *(German physicist, 1882–1970)*, 60

Boslough, John, 35, 52–53

Bossuet, Jacques *(French prelate, 1627–1704)*, 19, 25, 26, 27

Brahe, Tycho *(Danish astronomer, 1546–1601)*, 138

Calder, Nigel, 59, 60, 61–62

Capra, Fritjof, 52

Clark, Gordon H. *(American philosopher and theologian, 1902–1985)*, 94–95

Collins, James, 198

Copernicus, Nicolaus *(Polish astronomer, 1473–1543)*, 66, 69, 136–38, 139, 151

Copleston, Frederick, 173–74, 198

Darwin, Charles *(English naturalist, 1809–1882)*, 32, 163

Delbet, Pierre *(French biologist and surgeon, 1861–1957)*, 1, 8–11

Descartes, René *(French mathematician and philosopher, 1596–1650)*, 151, 190, 196–98, 202

Diderot, Denis *(French encyclopedist and philosopher, 1713–1784)*, 13

Index of
Foreign
Words

The foreign words in this index are Latin unless otherwise specified (Fr. = French, Germ. = German, Gk. = Greek).

Esse est percipi (To be is to be perceived), 94, 102

existere (to stand out of; to exist), 115

ex nihilo (out of nothing), 8

Ex nihilo nihil fit (Out of nothing, nothing comes), 8

gnōsis (Gk.: knowledge), 21

Grund und Folge (Germ.: ground [or reason] and consequent), 201

ignoramus (ignorant of), 21

locus (place), 199

Mea culpa (I am guilty), 48

modus ponendi (means of proposing, way of affirming), 177, 178

nomina (names), 25, 150

noumena (things apprehended by thought, independently of the senses), 203

organon (instrument), 30, 128

persona (an actor's mask, a character in a play; the person), 85, 86

posteriori. See *a posteriori*

Post hoc ergo propter hoc (After this, therefore because of this), 208

priori. See *a priori*

reductio ad absurdum (reduction to the absurd), 57, 58

res (thing, fact), 25, 150, 209

Sein (Germ.: being), 46

sensus divinitatis (sense [or awareness] of God), 124

sequi (to follow), 200

sic (so thus; written this way intentionally), 80, 115

sine qua non (without which not; something that is indispensable), 145

tabula rasa (blank tablet), 155

tertium quid (a third something; a middle course), 5, 54

totaliter Aliter (Germ.: wholly other), 107

tour de force (feat of strength or ingenuity), 7

Index of
Subjects

R. C. Sproul is a co-author of *Classical Apologetics* and has written and edited more than forty-five other volumes.

R. C. is founder and chairman of Ligonier Ministries, a teaching ministry that produces Christian educational materials designed to fill the gap between Sunday school and seminary. Beginning as a small study center in Ligonier, Pennsylvania, this ministry moved in 1984 to Orlando, Florida. With a staff of more than fifty people, Ligonier provides laypeople and pastors with substantive materials on theology, church history, Bible study, apologetics, and Christian ethics.

Ligonier's radio program, "Renewing Your Mind," features R. C. and is broadcast nationally, five days a week. Ligonier Ministries produces a monthly periodical, *Tabletalk,* has its own web site (see page 4 for the address), and sponsors several seminars a year, the largest of which is held in Orlando.

R. C. has taught hundreds of thousands of people through books, radio, audio and video tapes, seminars, sermons, seminary classes, and other forums. His goal is to help awaken as many people as possible to the holiness of God in all its fullness. His vision is that believers would apply truth to every sphere of their lives.

Dr. Sproul, a graduate of Westminster College, Pittsburgh Theological Seminary, and the Free University of Amsterdam, is professor of systematic theology and apologetics at Knox Theological Seminary in Fort Lauderdale and is ordained in the Presbyterian Church in America.